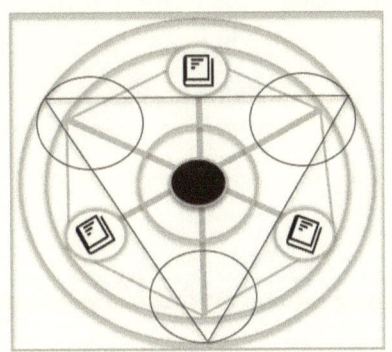

Alchemy of Tomes

Managing Technical Writers and Documentation Projects

Table of Contents

This book is dedicated to

D. A. Niel, for introducing me to new things, like Manga

Razi, for being an honest and wise voice

O. B. Won, for being the first to say, "I'm proud of you."

And to Annika Rubenstein, Master Alchemist

--

Acknowledgments

I must acknowledge Rachel Teitz, my true Beta Reader who read every line and dot. And D.O.D., who didn't really agree with my methods but helped read anyway. And in the end, I have to mention G.G.M., who kept bugging me to complete this. Thank you to everyone at SentinelOne who thought this was a cool project and didn't make me feel bad about using my energy for non-work-related things.

Special thank you to Steve and Anders and all the guys at Paligo - you rock!!

1. INTRODUCTION

"Alchemy's true definition encompasses the transformation of humanity to a higher stage"

—Israel Regardie[1]

"The first mission of a manager is to help her people improve."

—Annika Rubenstein

On a Monday evening after a typical hectic day, just as I pour a finger of my favorite golden whiskey, the phone rings. Caller ID shows the name of a former colleague.

"Hey! How are you doing? I haven't heard from you since we were both at company X."

"I'm doing great. We got our company going. I'll be managing the tech writer. I was thinking you could help me."

Was he asking me to join his start-up?

"What do you mean?"

"I don't even know what to look for when hiring. And can you tell me what I should know before I get into this?"

---

[1]Israel Regardie, *The Philosopher's Stone: Spiritual Alchemy, Psychology, and Ritual Magic*, (Llewellyn Publications, 2013).

"Are you asking me to tell you, in a phone call, how to manage technical writers?"

"Oh! No. You can write it in an email."

"My friend, I could a write a book."

Years later...

"Here it is...My book on managing technical writers!

"It helps writers and managers improve document projects.

"It has tips to improve quality, productivity, cost, scheduling, and more!

"It has templates for things like interviewing writers and issue reports!

"It has scripts for automatic formatting and formulas for automatic scheduling!"

"I call it: *Alchemy of Tomes!*"

"I don't understand the title. What is a tome?"

"A tome is a book, usually a heavy, old book, often in complex writing. It's ironic because technical writers do not write tomes. We write usable, cost-effective docs.

"And it's a play on words. 'Tome of alchemy' is a phrase, like: 'Frankenstein read tomes of alchemy to learn how to create his monster.'"

"What does alchemy have to do with anything?"

"Alchemy is about bringing something - or someone - to its best form and ability. A book about being a leader-manager is about being an alchemist of people."

"That is stupid. Why not just call it: How to Manage Technical Writers and Documentation Projects?"

"Did anyone ever tell you that you are no fun?"

"Yes. Often. Now give me the book. I'll read it after I finish updating my CV."

1.1. Why You Need this Book

Your technical writers can write traditional documentation that fulfills the product requirements. You can make sure the deliverables are done on time, with higher quality, with more relevance that reduces Support calls, and at a lower cost to the company. That is what this book is about.

This book is also for people like me – a technical writer turned manager. This book is a collection of my notes, tips, automation tricks, and lessons learned through (sometimes embarrassing) mistakes. I hope I can save you from the same trials!

Technical writing for corporations is a niche profession. This is not for copywriters, and marketing professionals might find only a part it interesting. But if you write or manage the documentation — manuals, knowledge base articles, product help, release notes — for software or hardware, you will find a lot here that is worth your time.

If you are the only writer in your organization, or if you are a member of a team with a team leader, look at the employee management sections as checklists. Which skills and characteristics do you have, and which can you improve?

If you are part of a team, please don't use this book to "educate" your manager. Encourage the ideas you like while letting your managers think it was their idea.

Sometimes this is called manipulative. I call it smart. Embrace your ability to influence others. It does not have to be a negative.

1.2. Notes about this Book

- I spent a lot of time searching for and trying different tools to create this book. I finally gave up on free and cheap tools. I asked my Paligo partner to set me up with a new instance. And now I am happy. **This book was created in and output from Paligo.** Check it out: paligo.net

- I wrote this on my off-time, usually while imbibing alcohol or too much caffeine. A careful reader will find mistakes. A sane reader will take what works and laugh off everything else. Please let me know if you find a mistake or my advice doesn't work for you. And let me know if you want (or do not want) to be quoted or acknowledged, and if so, how. You can write to: techpubmgr@gmail.com

This icon indicates a personal experience or explicit not-quite-serious note. A beta reader claimed these are the most telling parts. I will leave it up to you to decide.

- Use the same techpubmgr@gmail.com email to ask for downloadables of the templates, scripts, and spreadsheets that I show here.

You can send me $100 or point out a necessary correction for me to make. (If you are wondering, this is an example of my sense of humor.)

- A lot of this book is written in the imperative: a do-this, do-not-do-that, if-you-do-not-do-the-other-you-will-never-be-great type of sentence. I could say that this was the best way to make my ideas and opinions clear, but I am exactly what I seem to be. Let's call it assertive.

- Sentences sometimes begin with conjunctions. This is not an error. This is a conscious decision of style.

2. GLOSSARY

Command Line Interface (CLI)	Text-based user interface.
Component Content Management System (CCMS)	Software to organize re-usable topics with other files to create deliverables.
Content Management System (CMS)	Software to organize files of different types in deliverables.
Controlled Language (CL)	Subset of the natural language used to simplify understanding.
Curriculum Vitae (CV)	Used as synonym of *resume*. A document of a person's skills and education.
Graphical User Interface (GUI)	User interface with controls to click.
Human Resources (HR)	Corporate department that assists with recruitment, morale, and conflict resolution.
Localization (L10N)	Product changes for specific locales (translation, changes in colors, use cases, etc.)
Localization Manager (LM)	Manager of localization and translation projects.
Release Notes (RN)	Document of new features, fixed bugs, and newly discovered limitations.
Research and Development (R&D)	Corporate department of programmers, artists, and so on, who create the product.
Responsive Web Design (RWD)	HTML pages that refit according to screen size.
Return on Investment (ROI)	Measurement of whether a feature or tool is worthwhile by comparison of cost against expected results.
Subject Matter Expert (SME)	Person from whom the writer gains technical knowledge.

3. GETTING STARTED

 I'm making the budget for my department. Do I need a technical writer? If so, do we need a full-time, in-house writer, or should I look for a contractor? Into which group should I put this person?

3.1. What Does a Technical Writer Do?

The technical writer defines new concepts and explains how to use new technologies. The technical writer brings usable order to innovation chaos.

The output of technical writers can be in many different formats. They can be PDFs, internal procedures, GUI review, professional editing, XML files, video files, smartphone apps, augmented reality text, and formats not yet discovered.

With such a variable output, we must define the writer's responsibilities in general terms:

* Research the target audience and the company's evolving products and services
* Create instructions for users
* Review product interfaces for linguistic and usability errors

The mission of a writer is to create deliverables that enable people to optimize their use of the company's offerings. You can rewrite that statement to best fit your organization. It used to be:

> concise, comprehensive, consistent documents that enable rapid information retrieval during usage

I believe quality documentation is evolved to a different list of defining properties:

> controlled, relevant, accurate written self-service support

Controlled. If you control the language, it is concise. It is consistent between writers and for one writer over time. It is easier to understand, with better results[2], and cheaper to translate.

Relevant. If the product is basically intuitive, you do not need comprehensive documentation of every possible way to open a window or change a light bulb. But if you

---

[2]See https://www.linkedin.com/pulse/results-rochelle-fisher/ for the results of a test that proves this theory.

know your target audience well enough, you can create product deliverables that answer questions that users ask. These questions can be from "What can this product do for me?" to "How do I respond to this emergency?" to anything in between.

Accurate.   The content is based on real-world usage and tested on the product. It is updated for changes in the product releases.

The mission changes for each organization. The bottom line remains the same: read the user's mind and offer perfect assistance before it is requested.

3.2. Do You Need an In-House Writer?

That is a good question. But let's start with a more basic question:

Do you need a professional technical writer?

If you have other employees who can quickly turn blobs of knowledge into simple and accurate procedures, who can output multiple deliverables according to user expectations, who have the time in their schedules and priority in their expectations for documentation: then, no, you do not need a technical writer.

But asking your developers, testers, or support engineers to write the documentation is like asking your car mechanic to fix the kitchen sink. If you let the mechanic work on your car and hire a plumber for the kitchen, both jobs will be done faster, cheaper, and better. Hire a technical writer to write, and let the engineers do their work.

Let's say you need a technical writer. **Do you need a new hire, or should you out-source this effort?**

Consider these questions:

- How innovative is my product?
  If it takes time and expertise to understand your product, consider the time for training. If you can teach a contractor the fundamentals with an email or an hour presentation, hire the contractor. If your product is cutting-edge and constantly changing, it is more cost-effective to get a permanent, in-house writer.

- How often will I want a writer's input?
  If you want a writer to edit the posts of non-English speakers, to review the user interface while in development, or to document a continuously developing product, you want an in-house writer. Out-sourced contractors will not always be available to complete your edits on your timeline. With continuous work, a contractor will be more expensive than an in-house employee. If you expect product development to have a specific end time without significant changes for more than a quarter, a contractor can do that work.

- How can I reduce the costs of translation?
  If you develop hardware, many country regulations demand translation. To keep translation costs down, use consistent language. The more control your writers maintain over the language of the docs and the interface, the cheaper the translation. When you write according to a Controlled Language (CL), you must use an in-house team or a contractor with experience in and tools for that CL. If you use a contractor, you will have to use the same contractor for the next revision, or all the work and benefits of the CL are lost.

- Am I subject to regulations or standards?
  If your customers require compliance with government regulations, the documents are part of the full package. If you cannot find an out-source contractor with the required expertise and security clearance, consider a permanent in-house writer.

The bottom line: in-house or out-sourced, your writers require your guidance.

3.3. Where Does the Technical Writer Fit?

In a typical hi-tech company, writers are often first placed with Support. The writer and the support engineer answer the same questions. The problem is that they work on different versions of the product. The writers look for answers on how to use the version to be released. Support answers questions on versions that were released long ago.

If the writers are with Product Management, they will know the roadmap and priorities. But they will have to scramble for all technical information. By the time the writers are working, the Product Managers are already on the next features. The Product Managers will want to talk about the features they are focused on, even if the feature will not be integrated in the product for many weeks or quarters. If the tech writers do the research based on the internal documents and not on the reality of the product, they will waste a lot of time. First, the writers will have to do all the work again when the feature is implemented. It will change. Second, some of the best-planned features never get into the final product.

For the writers to be on the same timeline as the rest of their group, they could be in QA. The writer and the tester work on the same features at the same time in the product lifecycle. But QA does not have customer scenario information. They do the distinct test plans according to what the user is expected to do. But to get to a specific point, QA often uses internal workarounds that cannot be documented for the customer.

Sometimes the writers are with Professional Services. PS brings questions from potential users that help lead the documentation priorities for the next release. But PS usually does not have the technical answers to specific questions or information for general features.

In an agile environment, one writer can be with a feature team of developer and tester. In this scenario, your job as the manager is to bring the writers together. You will have to put in more time to make sure their results are consistent with each other and with the corporate message. You will have to integrate several feature documents in one product.

There is no right answer. Technical writers cover many departments, work with almost everyone in the company, and need and offer assistance from different levels and areas.

There is no right answer, but there are wrong answers. The biggest mistake is to put the writer in a miscellaneous department, such as Logistics or Operations, or under someone who wants to manage but whose field of expertise is too narrow to have a team. These scenarios generally isolate the writer. Research is more difficult, and relevant output is less likely. If this is you, make an effort to include your writers with the other teams. Make sure they are introduced to everyone. Make sure they know whom to contact, when, and how.

4. CHOOSING TOOLS

Understanding the tools of your team is a key ingredient in creating a roadmap that leads to your vision. The tools you choose must fit the vision you have for your team's results and methods. Thus, the tools are an important part of technical writing leadership.

You must know enough of the tools to be prepared to expand on them and to solve their limitations. Managers without technical writing experience are often inclined to let the writers choose their own tool. This is fine, **if you manage it**. Demand explanations. Ask the writers to test the tools and look for new versions of the tools they know. If they pick a free tool, be careful that it will give you what you need, when you need it. For example, if a writer insists he can work in Google Docs and deliver PDFs and HTML5, give him forty lashes with a wet noodle, and tell him to do some real investigation. Your tool must deliver professional output in minutes, not days.

4.1. Criteria to Choose a Tool

These are the criteria that I use. My team vision always depends on the writers working as a team, delivering quickly, and creating with quality. I learned through harsh experience to also think of tools that enable us to port our content to a different tool.

Things to consider when choosing a tool:

- **Portability** - Avoid technology that will lock you in the tool for years to come.
  For example, AuthorIt and Flare each implement a proprietary XML schema. Both tool providers continue their development, overtaking the other in one area or another with each release. But you cannot switch when the other provider has nailed the features you require. On the other hand, if you implement a tool that uses an open schema, such as DITA or DocBook, your writers can use the front-ends they like. And judge the tools for scaling up, not just sideways. If your team grows, will the tool still fit?

- **Teamwork** - If you have a team of writers, get a tool that allows for teamwork.
  Traditionally, tech writers worked as lone writers, even in a team. If a writer was out for some time, her projects were not done during that time. Or her projects were offloaded to another writer, who often started from scratch. Finding the latest version of a document can be tricky, even with strict guidelines for shared drives. If you use a Content Management System (CMS), it is much easier to hand-over, offload, and load-share as a team. If you have an international team (writers are physically in different locations), look for a cloud-based CMS.

- **Access** - Look for tools that let others (your managers, QA managers, and Support managers) have limited access to make quick, urgent changes.

If you can get this, make sure the tool also logs usernames and dates. You must review changes of non-writers and make sure they comply with the style guide and language. This idea can leave an unpleasant aftertaste in the mouths of traditional tech writers. If you have approval cycles in place, you do not want to let non-writers circumvent them. Think innovation, frAgile release, Agile development. A perfect offering two weeks late is obsolete in many industries. A notice of a known limitation given on time, even with grammatical errors, can save accounts. Just be sure to fix their writing as quickly as you can.

- **Centralized Variables** - Make sure the tool locks everyone to use the same variable in the same way.
  For example, standard Adobe FrameMaker, without the Adobe server, lets every user create their own variables. If you must rebrand many documents for one name change, it can take hundreds of hours. If you use a tool with centralized variables, rebranding is done in minutes. This becomes even more important if you send documents for translation. With centralized variables, each is translated once and delivered properly. Without it, you have a nightmare of overtime and too many fail-points.

- **Easy Interface** - *Easy* is relative. I find XML easier to write in than many front-end applications. Other writers need authoring tools with a graphical front-end. Know your team and choose a tool they can learn in the time you are able to give them to learn it.

- **Integration with Media** - Find a tool that delivers output in the media you want.
  If your main deliverable is PDF, do not choose a tool that creates PDFs from XML without an intermediate DOC file that you can change. If your main media is a Help Center, do not choose a tool that requires copy and paste to articles. There are plenty of tools that integrate directly to Help Centers. For example, Paligo delivers directly to Zendesk, and Author-It delivers filtered HTML5 for a request-response server.

- **Single Source** - This is a standard of professional authoring tools. Each publishes the same content to HTML and PDF. New tools or versions offer HTML5 and XML. Author-It and Flare can publish to digital readers (mobi, epub) with a little customization. What do you need today and what do you predict you will need in the future?

- **Diff and Revert** - Many tools offer levels of compare and revert.
  Even if your writers never make mistakes, products can change. The better you can see the exact difference between revisions of topics, the more prepared you will be for changes.

- **Metadata** - Tech writers must document from whom and when they got information.
  It is unacceptable for you to not know the answer when someone asks why a feature was removed or changed. If you have this information, it will prevent the common situation of one person requesting a revert, and another person demanding that the writer revert to a different point.

- **Status** - You must see the status of complete deliverables and each component in a deliverable at any time.

  Status can be a simple three-step choice of *Working, In Review, Approved.* Or it can be a more detailed list: *Assigned, Planned, Writing, Hands-on, Sent for Review R&D, Sent for Review QA, Implemented Comments, Approved.* If your deliverables are translated, or might be translated in the future, add status items: *Sent for Translation, Translation Review, Translated, Source Changed.*

4.2. Matrix Template for Tool PoC

I recommend that you track your Proof of Concept (PoC) trials of tools. I made a Google Sheet that lets me easily note the tools I test and compare the results according to my criteria.

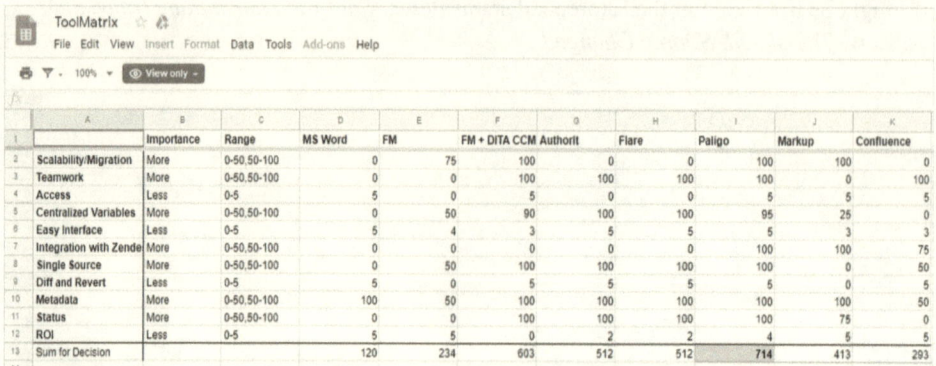

	Importance	Range	MS Word	FM	FM + DITA CCM Authorit	Flare	Paligo	Markup	Confluence
Scalability/Migration	More	0-50,50-100	0	75	100	0	0	100	100	0
Teamwork	More	0-50,50-100	0	0	100	100	100	100	0	100
Access	Less	0-5	5	0	5	0	0	5	5	5
Centralized Variables	More	0-50,50-100	0	50	90	100	100	95	25	0
Easy Interface	Less	0-5	5	4	3	5	5	5	3	3
Integration with Zende	More	0-50,50-100	0	0	0	0	0	100	100	75
Single Source	More	0-50,50-100	0	50	100	100	100	100	0	50
Diff and Revert	Less	0-5	5	0	5	5	5	5	0	5
Metadata	More	0-50,50-100	100	50	100	100	100	100	100	50
Status	More	0-50,50-100	0	0	100	100	100	100	75	0
ROI	Less	0-5	5	5	0	2	2	4	5	5
Sum for Decision			120	234	603	512	512	714	413	293

I have nothing against MS Word or Office, and I have no connection to Paligo sales or development. This sample shows my opinion, for the requirements of one company and its technical publications team.

To make your own matrix:

1. Enter your criteria in the first column.
   Notice that I added **ROI** and **Sum for Decision**. Add these too. I will explain how to use them in the steps below.

2. The second column header is **Importance**.

3. Select the cells in this column, excluding the header and the Sum row. Set Data Validation on these cells to be: **More** or **Less**.

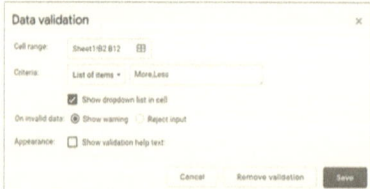

 I assume you know how to make a Data Validation list. If you do not know, open Google Docs or MS Excel and press F1.

4. The third column header is **Range**.

5. In the first cell under the **Range** header, enter this formula:

```
=IF(B2="More","0-50,50-100","0-5")
```

Copy the formula to the other cells in this column, excluding the header and the Sum row.

The formula means: if the value of **Importance** is **More**, show "0-50, 50-100". If it is not **More** (it is **Less**), show "0-5". This formula is a reminder, not a validation.

This gives you a weighted number for each criterion. For example, if the issue is important to you, for each tool give it a grade of 0 if the tool does not have that feature. If the tool offers it "somewhat", give it a grade of 50. If the tool handles that feature well, give a 100. When you compare multiple tools, you can use a range around those points to remember that one is better than the other.

If the issue is less important, the range is between 0 and 5.

The weighted grades will make it clear why one tool is off the consideration board, though it offers many features.

ROI (Return on Investment) is for the cost of the tool. Make sure to note the ROI of each tool in the grade range and not the real cost. A zero ROI grade means the tool is too expensive. You can keep the real cost in a note of the cell, or add rows for the vendor contact details and the cost. I do not show it here, but keep the contact information with the name of the tool. You want a record of who talked to you and how to get in touch again.

6. The fourth column is for your first tool. In the Header row, enter the name of the tool.

7. In the Sum row of the first tool, enter the formula to calculate the sum of your weighted grades:

```
=sum(D2:D12)
```

Copy the formula to the Sum of each column of the tools you will look at.

I recommend that you set the Sum cells to conditional formatting, to highlight the highest grade. When you test five or ten tools, you want to make it as easy as possible to show the budget makers why you chose the tool you did.

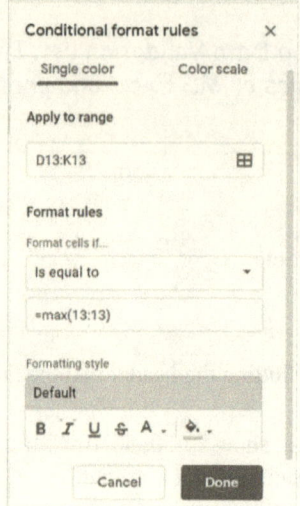

5. MANAGING DOCUMENTATION PROJECTS

I've heard this so many times, just thinking about it again makes me want to scream:

"You can have it fast. You can have it cheap. You can have it done well. But you can't have all three."

Bull.

You do not need it done quickly. You need it done on time. You do not need it done cheaply. You need it done with the available resources. You do not need it to be perfect. You need it to reach the success criteria set in the beginning by the project owner.

Reaching quality output, on time, without overtime, requires good planning, follow through on your part, focus on the part of the writers, and openness between you.

5.1. Basic Truths

Project Management is basic to all types of development. But if you have experience as a project manager, do not skip this section. There are truths about documentation projects that will change your management strategy.

5.1.1. Time is Different Between the Worlds of R&D and Documentation

Know this first truth, O Grasshopper: The time it takes to develop a feature has NOTHING to do with the time it takes to document it.

Technical writers document what a user must do to reach an objective.

We are not talking about White Papers. That is a creature in its own right. If you are expected to create White Papers, learn the standard outline and characteristics. And accept my condolences.

If a feature does not allow for user interaction, there is nothing to document, other than a line in the Release Notes.

For example, if R&D spends six months changing the product for better performance, what is there to document except two words in the Release Notes: *Improved performance.* If your organization has a performance testing team, maybe you can write a paragraph on ratios and bandwidths. It is more likely you will edit that paragraph from R&D.

But if R&D takes six minutes to replace one word in the interface with a different word, your team has hours or days of work. The GUI label must be changed everywhere in the text and in the screenshots. If you deliver videos, all the videos must be redone.

5.1.2. Ownership: Possession is Nine-Tenths of Crazy

In the 1990s, ownership was the hot topic of the newly blooming tech writing field. There were tech writers all along, but as a massive new opportunity, it was just beginning.

The question of ownership became an issue when developers and upper management started seeing the writer as a member of the family, rather than an outsider. Writers started delivering PDFs. Compared to doc or text files, PDFs were engraved in stone. No one could change your writing. All the developers could do was comment on things they wanted to change. This was important to writers. We didn't want to sign off on printed books with changes we didn't control.

With the influx of CCMS documentation databases, this got out of hand. Now we couldn't make even a small change without publishing the complete document from the source. Simple fixes to the output were undone on the next publish from the source or were put off for weeks.

Let's look at this question one more time. Who owns the documentation? I'm going to give you my subjective opinion.

And no other opinion. This is my book, and I can laugh if I want to.

The company owns the documentation. If the CEO or department head wants to scrap a project you worked on intensively, scrap it. If they want to change the name of a project, you change it. If they want to change a common noun to a new proper noun, you find a way to change it - carefully! Make sure the indefinite articles (a versus an) match the new name.

You are responsible for the quality and deadline of the document. You must sign off on a project, as approved, as fulfilling the definition of *done*.

In the second and third decade of the 21st century, we find ourselves in the age of the FrAgile Release. New ideas are brought to market before the product is complete. New products are pushed up in release, to meet competitive demands. Through all this high-intensity Agile cycle of develop > release > fix > release > develop again, you are responsible for the documentation. You never hold up a release for documentation, and you never allow a release to go out without some document.

You do not have the last word in what goes in or whether it should be changed. The documentation is part of the product package. The project sponsor has the last word.

5.2. Tracking

How long will it take for one writer to research and write the first draft of a feature and its procedures?

As a manager, you must be able to make accurate estimates for complete projects. You must know how long it takes to document types of features, create diagrams, take screenshots, proofread, implement comments, and deliver. You will be asked to judge your writers' productivity and compare it to expectations. The best way to get these answers: keep a Time Tracker.

5.2.1. Communicating Tracker WIIFY

When it comes to overhead tasks, it is important that your writers understand the What's In It For You (WIIFY) message.

Concepts to communicate to your team:

- Tracking makes it easy to estimate the time needed to complete your tasks.
  Often we under-estimate how long it takes to do the tasks that we know how to do. If we track our activities, we can easily pull up that number of required hours. This means no over-time for you to complete a task for an incorrect due-date.

- Tracking helps us focus on areas of improvement.
  If you think that you spend too much time on making network diagrams, for example, we can compare the time you spend on a diagram with what others spend. Maybe it is a time-consuming task, and we need to prepare for that in plans. Or maybe someone on the team is doing a better job. Maybe they want to do all the diagrams. Or maybe they have tricks to teach the rest of us. Or maybe there is no need to improve the process. Tedious or difficult tasks seem to take longer. We need to track to find out.

- Accurate tracking lets me make accurate plans.
  When they say a new service pack (SP) will be released in two months, I might think that's not a problem for us. But when I look at the tracking, I see that the last SP took 800 hours for complete documentation. From the beginning, I can ask for an Impact Plan, to cut off doc requirements to fit our time. This means no end-of-release rush time.

- Accurate plans let me prove headcount needs.
  Knowing the required time to complete a release lets me ask for more people when the company makes a new product line. Without accurate reports on resource value, we will have a much harder time getting the people we need.

5.2.2. Choosing a Tracking Tool

Some organizations have a proprietary or paid-for tracking system. Employees log each activity when they start and finish. If your organization uses this, make sure your writers understand why this is important, and which activities are important to log.

There are many free tools. I have tried some, but I found each lacking. One we used a lot was an online WebUI. But a Web session would end, and the form to submit a tracked activity would still show. After submitting all the data, we would get the time-out message and the data would be lost. We also had problems with the reports. The tool accepted all time and date formats, but the reports did not convert to one consistent format. That was in 2009 or 2010. I'm sure technology has advanced. Maybe you can find a free tool that works for you.

My favorite is a spreadsheet – MS Excel or Google Sheets. This solution depends on the honor system. If you come in to manage an existing team, keep a watch out for manipulations and delinquent reporting.

5.2.3. Creating a Spreadsheet Time Tracker

If you use MS Excel, it must be a version that supports Power Pivot. If you try to use one shared spreadsheet, writers will quickly become frustrated when they are locked out. If each writer uses their own copy, you must make sure they all use the same formats and columns. Then you create pivot tables to combine them for a team view. Without Power Pivot, your reports will take you hours to create.

If you use Google Sheets, make sure you do not overwrite each other. If one writer is on a row, the next writer should take the next row.

With all spreadsheets:

- Make sure there are no empty rows. That will ruin your pivot tables. If a column is optional and empty in some rows, you cannot pivot on that column's data. If a complete row is empty, the pivot table ends there.

- Plan when to start a new Tracker file. I suggest every quarter for a team of 3 on one Google Sheet, or every half year for a team of up to 10 on local Excel files. If you continue a file too long, it will be too heavy to use easily.

- Learn how to make pivot tables, to summarize closed trackers, or how to use the Google Sheets QUERY function. The Google Sheets EXPLORE function is also nice and often all you need. With a large team, you can learn Python or ask a developer for help. Python comes with spreadsheet libraries and a collection of libraries for big data. After the script is created, you will get your estimates in seconds.

- Keep all original trackers. They are often necessary.

- Use Data Validation lists as much as possible. If writers enter values, rather than select from a list, you will spend too much time trying to organize similar values that should have been identical.

To make a spreadsheet Time Tracker:

1. Make one tab called **TimeTracker**.

2. Make one tab called **Values**.

3. In the **Values** tab, enter the values to select in the tracker. These are the columns of values I use. Make sure each cell has only one value.
    Writer – List of usernames or initials of your team.

You would be surprised to know how many hours I spent fixing pivot tables because writers misspelled their own names or used inconsistent formats.

Activity – List of activities that your team does. Some writers like to have a complete list that includes items such as Reading Emails, Writing Emails, Meetings, Phone Calls. Others like a shortlist of product management phases: Initiation, Planning, Writing, Closure. You will know what you need when you make a report for your managers or work with a writer to improve productivity. I suggest you work with the team to tune this list. I like a list of about ten.

- Planning
- Researching
- Writing
- Graphics
- Videos
- Editing
- Peer Reviewing
- Implementing comments
- Delivering
- Training
- Translation

We generally overestimate the time required for translation. Preparing, importing, and delivering translation can be tedious for a technical writer. It is important to track exactly how much time it takes.

DocProject – List of your deliverables. Even if you single-source, there are different types of work for different projects. These are the values I usually use.

UG	User Guide, Admin Guide, Technical Guide, in all formats (PDF, HTML, HTML5, DOC, TXT, man page, etc.)
RN	Release Notes
CL	Controlled Language (edit writing for compliance, work on the vocabulary or rules)
HC	Help Center (edit Support articles and Marcom blogs)
UI	Product User Interface (create bug reports for language errors or usability suggestions in the GUI or CLI)
Infrastructure	Templates, tools, methodologies

ProductRelease – List of the Product Manager milestones for which you supply documentation. This is usually the release name of the product.

Status - List of status, such as Planned, In Progress, Pending Issue, Pending SME, Pending Approval, Done.

4. In the **Tracker** tab, enter these column headers:

Column Headers	Description
Writer	Writers select their name from the Data Validation of the Values sheet > Writer column.
Date	One specific format. (Tip: If you use a format that Google Sheets recognizes, the writer can click in the cell to open a mini calendar and then click the date.) Select the column and set the format of cells for the date format you want everyone to use.

Column Headers	Description
Activity, DocProject, ProductRelease	Data Validation lists, from columns in the **Values** tab.
Duration (hrs)	Decimals of hours. For example, one and a half hours is not 1:30, but 1.5. I find the decimals easier to see incorrect calculations, easier to convert to days, and easier to enter. Do not set Data Validation on this column. If you use your TimeTracker as your Weekly Goal Planner, this value can be *Goal*.
Feature	This can be the object code in a CCMS, a user-defined word in a new Data Validation list, or a generated key. This lets you see the duration of one thing – one feature, topic, graphic, video, or report. With this data, you can find Min and Max or Average, to set team expectations and individual areas of improvement. The writer uses the same value for each row that gives a duration for the one thing. For example, if a writer works on a feature over five days, it is usually not five full days on only one feature. We work on many features and tasks every day. The feature shows the sum for one thing, over time.
Goal and **Issue**	If you want to keep your Weekly Goals with your Time Tracker, add these columns. Goal is a due date. Issue is text that explains why a goal was not met. (Tip: At first, Issue is free text. Review the issues, solve as many as you can, and then make a list of 5 to 10 categories that cover the issues your writers deal with. Put that list in the Values tab. Then, use Data Validation for the Issue column to use that list.)
Status	Data Validation list of status.
Notes	Free text. If you do not set aside a place for notes, your writers will ruin your Tracker, trying to put notes in other ways.

Power Tip for Google Sheets: Even one person's Tracker can quickly become long. Add this script to open the sheet to the first empty row:

```
function onOpen() {
  var ThisSheet =
SpreadsheetApp.getActiveSpreadsheet();
  var StartingTab =
ThisSheet.getSheetByName("Tracker");
  var LastRow =
ThisSheet.getLastRow() + 1;
  var LastCellString =
"A"+LastRow+":A"+LastRow;
  var LastCell =
ThisSheet.getRange(LastCellString);
StartingTab.setActiveRange(LastCell)
}
```

5.2.4. Time Tracker Example

Writer	Date	Activity	DocProject	ProductRelease Feature	Duration (hrs)	Notes	Goal deadline	Issue
AF	2018-12-16	Writing	UG	2.8 Win	6		2019-01-24 Done	
AF	2018-12-16	Writing	Training-LMS	PreSalesCou	done		2019-02-07 Done	
CH	2018-12-18	Writing	UG	Prj X	Goal		2018-12-26 Pending Issu	
BG	2018-12-30	Delivering	RN	Prj X	Goal		2019-01-03 Done	
BG	2018-12-30	Delivering	UG	Prj Y SP4	Goal		2019-01-03 Done	
BG	2018-12-30	Videos	UG	Prj Y SP4	Goal		2019-01-03 Done	
CH	2019-01-01	Writing	Training-LMS	all		2		

If a goal is completed in one continuous effort, the writer can replace `Goal` with the duration. (It is still recognized as a goal because it has a due date.) For this reason, the Weekly Goal in the Time Tracker is better than in another tool: less overhead, all in one place.

Let's look at part of the Time Tracker of some of my writers: Alex, Betty, and Charlie.

Alex had a goal to write a feature for a guide for version 2.8. At the beginning of the week, he entered `Goal` in the **Duration** column. When he completed the guide in six hours, he replaced `Goal` with 6. In the **Status** column, he selected **Done**.

Charlie had a goal to write a feature for the Project X guide. Charlie did not complete the goal on time, so selected **Pending Issue** in the **Status** column and briefly explained the problem in the **Issue** column.

Betty had a goal to deliver the Release Notes for Project X. She tracked her work during the week. Each time she worked on this goal, she entered the duration. On Monday she worked on for an hour, on Wednesday, for two hours, and so on. When she completed the goal, she selected **Done** in the **Status** column of the goal row.

We often use the Google Sheets Explore feature. A typical question you will ask in Explore is: "Sum of **Duration** for **ProductRelease** of *product name*". This gives the sum of hours that the team worked on a product. On Excel, you can create a pivot table to get the same answer.

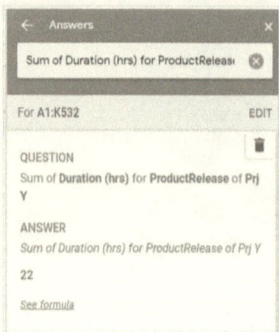

"We have a new release of Product A. The release is called X. How long will it take your team to complete?"

"As a project, is it similar to Y?"

"Yes. It has a similar scope of features."

In seconds, you can run Google Sheets Explore and see that Project Y took 22 hours. To change hours to working days, divide by 8 hours a day and round up. Thus Project Y took 22 divided by 8 = 2.75 > 3 days.

"It will take us 3 working days to do our part. We need to have these days split, to leave room for SMEs to review. We will have to remember to request that the release stakeholders take into account time required to review documentation."

"Project X adds enhancements without GUI. You will not need to make new video tutorials."

Sum of Duration (hrs) for ProductRelease of Prj Y and Activity of Videos = 14 > 22-14=8 > 1 day.

"Then we can complete Project X in ...1 day. I expect maybe half a day for writing and half to implement comments from reviewers."

5.3. Documentation Lifecycle

This is an example of the complete lifecycle of a document. Adjust it to meet the culture of your company and the requirements of the documentation. For example, this lifecycle shows that the writers get approval at every stage. If that does not fit your company culture, change it.

Preparation:

1. Get a request for a new document.
   This can be from the R&D roadmap, bug reports, future issues, or a documentation request form that you create for users or employees to fill and submit.

2. Make a plan. Use the team documentation project plan template.
   Collect all available information and specs to understand as much about the document content and requirements as possible.

3. Meet with the request submitter or project manager.
   Come prepared with specific questions and missing information for your plan.

4. Update the plan with deadlines for drafts, final release, scope, estimated effort, and priority of the project as a whole.

5. Discuss the schedule with your manager.

6. Create the outline and get approval.

7. Prioritize features of the product.
   If the schedule changes, or there are other risks, what features can be dropped or postponed?

8. Prioritize features of the document.
   Do you need wire-frame pictures, screenshots, videos, voice-overs, index, help center labels, and so on?

9. Update the schedule and plan.

Creation:

1. Create the structure for the document.
   In a CCMS, this step is to create the container object (book or publication). Get the unique part number, download ID, and similar document project data.

2. Create feature content. Write according to the style guide.

3. Proofread.

4. Send it for peer review to another writer.

5. Implement comments, metadoc questions, and changes. Update the status of the topics.

6. Send it for approval to the R&D SME and to the QA SME.

7. Implement comments, metadoc questions, and changes. Update the status of the topics.

8. When all features are written and approved, or it is the deadline for the first draft and it is good enough, send the draft to the draft reviewer (the project owner, according to the plan).

9. Implement comments and send the final draft to the project owner or sponsor.

Delivery:

1. Publish or compile to the output formats. If upload is a separate step, follow the procedures for upload.

2. If you deliver a printed book, prepare the document for printing.

3. Commit the output for storage as required by the company policy.

4. Close the project according to company procedures.

5.3.1. Milestones - Get the When

What is the release date? Your due date is the release date minus three days for final review.

Three is a magic number, so they say. It seems to work for most organizations.

The three-day rule for review and changes is true for documentation that requires more than 75 hours of work.

If you can deliver a Release Notes document in one or two days, and your SMEs and the project owner work with you, you can start and complete the whole thing two days before the release date.

If your SMEs are less responsive and the project is bigger, your due day is the release date minus seven days or more.

If the project is continuous (for example, a cloud application that will be updated monthly or an Agile project with sprints every two weeks), plan for the future due dates too.

A high-priority, one-time project that will take less than one day does not require special management. Assign it and make sure it is delivered. Everything else requires a schedule.

5.3.2. Document Request Form Template

Project Details:

Project internal name & Release name/number:

Purpose (Driver and priority of release, priority of its docs, as a first-line of defense against trade-offs):

Query targets:

How to get the product for hands-on:

Target audience and assumed knowledge:

Milestones:

EA date:____ GA date:____Other dates:_____

Stakeholders:

Doc request submitter:_____ Project sponsor:_____

R&D owner:_____ QA owner:_____

Release Mgr:_____ Final approver:_____

Features:

Link to PRD, wiki, other scope descriptions:

Localization requirements:

LM: _____ Languages: _____ Reviewers: _____

Deliverables: (check required and prioritize)

☐ Guides (Admin, Installation, User, Troubleshooting, Getting Started); PDF, Help Center, HTML, HTML5

☐ Release Notes

☐ F1 Help

☐ Videos, eLearning

5.3.3. Find the Designated Driver

What is the purpose of the documentation for this project? Start with a definition of the target audience: when and how frequently do you expect the users to read the documents? What knowledge must the writers assume the readers already have?

Make sure you get the answer from the stakeholders: the managers, SMEs, and other people who have a stake in the success of the project.

It is important that you not accept a lazy answer to the question of assumed knowledge. If the document explains concepts that the users must already know, they will think the documents are worthless before they get to the unique procedures. If the document explains only how to use the product without any conceptual explanations, you might lose users who do not understand why or when to run the steps you give them.

What are the most important drivers of the release?

If the project must go out on a specific date, with the available technical writers you have now, find out if you can complete the scope with those working hours.

```
=8 hours x Number of working days until due date x Number of
writers
```

If you cannot make the due date for the full scope, give the project manager an impact plan. This is a typical email to start the discussion:

This is what we can do by that date, ordered by priority: Feature A, Feature B, and Feature C. What do you want us to do about Features D and E? They will take ten more days. These are the options I suggest:

1. Give Features D and E in a second phase on GA date + 10 days.

2. Drop Features D and E completely from the documents.

3. Postpone these documentation features for a second phase: screenshots, videos, graphics. (That will give us 12 more days.)

Can we discuss this today?

Maybe the project owner will decide to give up on a different driver. If the date is more important than the budget, you might be asked to hire a temp to help. You must be able to estimate the number of missing hours. Those hours must be for tasks that a temp without training can do.

5.3.4. Priority by Purpose

What is the purpose of the project, in the offerings of the company?

You are invited to a kick-off meeting for a new product. The manager explains that the product is a new side utility. He expects R&D to complete it in one month. At that point, he will present it to the VP of Products. For this manager, this project is the question of whether or not he continues his career in this organization. But your team is committed to higher priority products. From the beginning, you know you will not get trade-off agreements from the other project owners to work on this product. It has no sales, no market, and no priority (yet) for the company. Make a note to get back to the project owner on the side (not in the meeting) to offer different options.

During the meeting, understand the product manager's drivers. Which of these things is most important for this project: time (delivery of milestones), resources (people and money), or quality (based on the definition of quality for this project)?

- If budget is not a problem, it would be wise to suggest that the project owner get a budget to hire a contractor for one month to document this one project.

- If the driver is time, and quality is not an issue, suggest that the project owner go to the presentation without documentation. Come back when the project has priority.

- If documentation is a requirement for the quality of the project, but quality is not defined for the documentation itself, suggest that R&D document the product. You can give it one or two days to edit.

If your team is open for more work: plan, assign, and do this project as though it had priority.

If a project is a normal evolution release, you still require the purpose. You must understand the main purpose to prioritize features, suggest trade-offs to handle schedule conflicts, and create a relevant impact plan if the schedule or scope changes.

5.3.5. Use a Scope to See Further

There are different scopes in a documentation project: deliverables, product features, and documentation features.

1. Discover the required deliverables before you start.

- If the project owner wants a series of help desk articles, do not plan for PDFs.
- Make sure you know the full expected list of deliverables. Does the project owner expect Release Notes, Admin Guide, and Installation Guide?
- If the purpose of the release is to supply bug fixes in a patch, you can suggest that the documentation be only a Release Notes PDF or HTML article.
- If the purpose of the release is a cloud application, the scope of deliverables depends on the target audience. If the target audience is users of the application, your deliverable scope is only a User Guide and Release Notes. If the audience is Sys Admins, DevOps, or others who handle the back-end of the cloud, your deliverable scope is much more complex. You will need an Installation Guide, a Deployment Guide, an Admin Guide, Interoperability Guide, Sizing Guide, Troubleshooting Guide, and Release Notes.

2. Discover the required features of the product.

- Product features that are directly connected to the primary purpose of the release are your priority 1 features.
- Of the secondary features, if there are some that change existing documentation, they are your priority 2. For example, if the product has a lower-priority feature to change the GUI, it is a medium feature for you.
- If you have bugs or missing information from previous versions, their priority in your plans is according to the priority of the product manager. You'll have to discuss this after the kick-off.
- The remaining product features are your lowest priority.
- If a development team consistently adds features after the planning stage, plan for added features in the scope. If your plan is too tight, you can move *unplanned features* to the Risk Management Plan.

3. Discover the required features of the documentation. Look at the success criteria of the product to help decide this scope.

- If quality is the driver, and the audience is the end-user, screenshots for every piece of the GUI is a priority for every feature.[3]
  Remember that screenshots require future maintenance. Each shot added is a documentation feature in itself. If the GUI changes in the next version, do you have the resources to update the screenshots? Some companies give free rein to the UX and front-end teams. You might not know of a change until someone opens a bug for your team.
- If time is the driver, and the audience has a large assumed knowledge, screenshots are your lowest priority.

---

[3]This is especially true for the Japanese market.

- Other documentation features are: glossary, graphics, tables, lists, text-based user interfaces (CLI, API, SDK), printed books, and videos. You might have a different list. The idea is that your schedule is based on the things that you deliver, not only the product feature list.

Count the number of screenshots that are required to add or change. Count the number of graphics (such as network topology) that are required to add or change. When you have some tracking, you will be able to estimate how many hours are required for the screenshots and graphics.

Remember to add to the scope documentation features that you always require. For example, does your documentation always have an Overview and a Glossary? Do you think they should be there? The more granular you can go, the more flexible your plan can be.

Make sure your definition of *DONE* fits the product manager's expectations and your schedule.

"Hi, Robert. I see the Success Criteria for Project X is a higher rating from an external Usability Testing company. Part of that rating is usability of the documents."

"That's right. We want to improve usability across the board."

"We can offer more documentation features. The last rating complained that users had to login to an external site to download the PDFs. Can we change that?"

"We have to maintain security of the docs, so there must be authentication."

"We could deliver the documents as articles in the knowledge base. That would let users run the full-text search without a download."

"If you can deliver that on time, it sounds like a good plan. What other ideas do you have?

"We can add more graphics for the introduction, where we explain the architecture. We will see if the UX designers have time for that. We can also add tutorial videos to the knowledge base."

"I would rather you did not take time from UX. We need them for the GUI. But videos sound like a great idea. How many can you make?"

"I will get back to you with a list of videos and you can prioritize them. On average, it takes an hour to make a minute of video. I have to work out our complete schedule before I can commit to a number of videos."

"Sounds good. Let's meet again in a couple of days."

5.3.6. Plan for the Unplanned

There are many tasks that a technical writer does for different projects. Keep these tasks in mind when you make your plans.

- GUI labels, error messages, log files - Even with a UX team, the technical writers are often required to review labels for correct grammar

- Code comments
- Marketing Requirements (MRD) - summary of product requirements from market research or by specific customer request
- Requirement Definition (RD) - summary of product requirements, especially those promised to customers
- External Product Specification (EPS) - product details and features
- High Level Design (HLD) - how product requirements are met in the design
- Internal Product Specification (IPS) - detailed technical implementation plans based on the EPS
- Product Implementation Plan (PIP) - development milestones for managers
- Test Plan - items and methods for QA

5.3.7. Expand the Scope for Translation

Make sure you are aware of localization requirements. If the documents are expected to be translated, you must enter that in the plan. Even with a Localization Manager and Translation company, your writers will have some work to do. If the product is completely localized – GUI translated and maybe even changed for color or layout, Help translated, docs translated – you must add at least a month for the translation to begin. If the product must be delivered on the milestone in all languages, your milestone is one month before release.

If screenshots are required, add up the time needed for all the shots in the original language, multiplied by one-third, multiplied for each language, to understand the greater effort to take screenshots in a language you cannot read. Then add one day for each localization to get access a translated GUI. Round up to make a full day estimate.

For example:

You expect English (source language) screenshots to be done in 3 days. The GUI will be translated to FIGS (French, Italian, German, and Spanish). This is the total number of days you need to get screenshots in all the localizations:

$3 + ((3 * 1.67) * 4) + (1 * 4) = 27$ days required for localized screenshots.

Note – If the product manager agrees this is too long, and one of the languages is Japanese, the Japanese document must include screenshots if the original has the screenshots. The organization's reseller in Japan might be able to take them for you.

5.3.8. Analyzing Target Audience for Documentation Features

You must know what you can assume the users know and what you can assume they expect in the documentation. If your audience is system administrators for the Installation Guide, and one of the features is support for Two-Factor Authentication, do not waste documentation space explaining what 2FA is and how to get it. The sys admin already knows. The documentation must include a procedure to set up 2FA for end-users with your product, without relating to the different vendors.

The only reason to have your writers learn back-end technology of assumed knowledge, is to write troubleshooting. If the project has MTU hop configuration, your writers do not spend hours learning MTU history and usage. They can learn if they are interested, but they must not explain it in documentation. They can document troubleshooting. This leads them to ask questions that most developers cannot answer: What issues are solved if the user changes the MTU? What are the best practices or troubleshooting workflows?

If your writers speak only with developers, the documents will be too simple: "Enter an integer in the MTU." "Enter the 2FA code from an authenticator."

If the documentation explains everything the writers learned, the documents will be too complex. The documentation will be pages of technology explanations that must be second-rate to the textbooks that the users already studied. The worst part of this issue, is that the readers will skip over all of it, lose the information they really need, and lose respect for your deliverable as a relevant tool.

Your writers must speak with the people who work with the tool or who work with users: Customer Success, Professional Services, Support, and QA. Experience with users is the best basis for audience analysis.

5.3.9. From Model to Dates

With the scope, audience, and milestones, you can begin to plan the work. The first planning stage should take about 10% of the total time for the project. If you take longer than 20%, you are wasting your time. By the time you get to work, the product will have changed, and your strict plans will be worthless.

Choose Your Weapon

The documentation plan depends on the development model you choose. These models work for documentation, from my experience:

Spiral: Work on topics as separate components. Send the topics for review and change each piece. Put the topics together in the final deliverable when each has reached a state that fits the success criteria and acceptable quality definition of the product manager.

Waterfall: Outline > {Research > Write} > {Test > Fix} > Send for review > Fix > Deliver. If you choose this route, make sure that writers use mini-waterfalls. Do not let them plan to learn everything and then write. This leaves you open to a huge management failure: one person takes scheduled time to learn a product and then becomes unavailable. Now you must duplicate that effort. The efficient plan is to create a complete outline, research a feature, write it, test it, fix it, and then go on to the next feature.

Prioritize by Importance and Effort

Start the project plan according to the model you choose.

1. Create a list of product features and documentation features.

2. Give each a priority number.

3. Use your Time Tracker to find the estimated effort for each feature.

When you plan an employee's time, NEVER define a day as nine hours. Consider the humanity of yourself and of your people.

The only way to produce nine hours of work in one day is to work more than ten hours.

This information is the beginning of a complex, multi-writer plan. But you can also use this to help one writer. When we expose too much of the big picture to our writers, they can become overwhelmed. If this happens, your writer should come to you with, "I don't know what to do first." Help your writer create this table. I like to use a whiteboard. It makes it easier to change as you go until you have the solution.

The writer enters the list of tasks to do, in any order. Then, the writer guesses at whether the task will be difficult or easy. The easier, the lower the number. You can use hour estimates here, if you can do it from rote. Enter the due date. This is the date that SMEs committed to review or, if you didn't get commitment, the release date. In this temporary, whiteboard table, we are setting priority, not schedule.

In this example, Betty came with a list of tasks.

- Create VMs for team to take screenshots of different operating systems
- Learn and write Feature B
- Peer review for the topics that Alex wrote, for Features A and C
- Write Release Notes for Project X

We realized that she must complete the Release Notes due the next day before she does anything else. We set that to Priority 1. Then, Betty explained that she must create the virtual machines for the team to use to document a new product. Everyone was waiting on this. That became her priority 2. Now she has two tasks, both due later. We gave a higher priority for the easier task.

Pri	Task	Difficulty	Due
2	Create VMs	5	Thursday
4	Learn & write Feature B	2	2 weeks
3	Peer review for A & C	1	2 weeks
1	Proj X RN	4	tomorrow

When Everything is Priority One

Often you will ask for priorities on features, and the project owner will say everything is important. Show your schedule to do all the product and features, with a thick red line after the release date. Add a temporary task for closing a partial project: proofreading, one review and one fix, delivering.

This is the Impact Plan. Use it in the beginning, when features are added after the first plan, or other projects take your team's time. Only list the things you can control - the features you know of and the quality you expect. You can't show the impact of your plan if you plan from the start for unknown features and issues.

"This above the red line is what we can do before the due date. We can drop everything below. We can deliver in phases. Or we can rework the plan, so that lower priority features are below the line. What do you want to do?"

When you put the project owner on the spot like this, take notes on paper, not a laptop or PC. You don't want anything between you two. This body language reinforces the concept of teamwork and equal stakeholders in success.

Note about more people: If the project duration is less than a quarter, getting a temp is usually not effective. They need time to learn your tools. If you have a number of projects that your team cannot complete, track that situation. It is time to expand the team.

5.3.10. Defined Quality

Define success criteria of the project. What is quality? What does *done* mean? For example, you can define Quality as 100% proofread by at least 2 writers (the lead writer and a peer or outside editor) and approved by QA and R&D for accuracy.

Make sure the plan includes time and resources to meet your success criteria. In the example, you would need to add time for proofreading in the project, schedule of the proofreader, and time for review-comment implementation. At this point, it will be clear that you also need a commitment from QA and R&D to review the guides. You will have a good idea of the dates when they will get topics to review. You can ask for a commitment to get comments back by specific dates.

5.4. High-Level Schedule

The high-level schedule, or Master Plan,

MmawAHAHAHAHAHAahahaaa

Evil laugh is required.

shows when releases are due, and it shows which days you will work on them.

When you first begin to fill in the days for a new project, you will probably put in too much. Update it as the writers get closer to realistic efforts. If you do not keep a realistic schedule, you will ask for trade-offs when none are needed.

When priorities change, do not replace one project with another. Move the first project further up the writer's schedule, or move it to an open space on another writer.

The schedule can be a Gantt chart, or a calendar (horizontal or monthly representation of the days of the year), or whatever visualization of dates for multiple people floats your boat. The Weekly Goals must fit the schedule. The schedule shows you clearly if you are on the road to meeting your vision, if you need more resources, and when you have an opening for a new project.

5.4.1. Schedule Trade-Offs

Consider your organization as one entity, with one purpose. Each project for each product is one requirement in your documentation product. When project schedules conflict, handle them with trade-offs, in the same way that a project manager handles requirement trade-offs. When you have plans for each person on your team, and a new request comes in (a new product or a change in a scheduled project that requires more resources), do not answer yes or no immediately. If you must respond, do it with, "I will get back to you as soon as I can."

To handle a trade-off:

1. Gather information. What is the scope of the new request? When is it due? How many hours or days will it require?

2. Draw a timeline with writers and their projects: milestones, number of days before each milestone that you must lock in for proofreading and implementing major comments, high-priority features that must be documented by the milestone, lower

priority features that can be added after the milestone, and documentation features (graphics, screenshots, videos, glossary, etc.) that you can drop or delay.

3.   Mix and match. Match the writers who have expertise with the projects they can do faster than others. Ask writers to take over features from other projects.

Here is an example of data collection for a trade-off and the solution we discovered:

	Project X	Project Y
Milestone:	EA. Feb 28 GA. Mar 15	LA: Jan 15 GA: Feb 15, Mar 31
Writer	Alex	Betty
Remaining features and effort	5 easy (3d), 5 hard (25d). 25 shots (5d)	9 easy (9d), 1 hard (5d). 10 shots (2d). F1 mapping (avg 2d)
Commitment for review	project owner committed to complete review at the end.	3 teams of QA and R&D reviewed topics. 10 new topics to do.
Lower priority documentation features	Graphics 6d Videos 5d Scenario examples 1d	Graphics 1d Videos 5d CLI Alternatives 3d
Required documentation features	Peer review 2d Proofing 1d Implement comments 1d	Peer review 1d Proofing 1d Implement comments 1d

The company plans to release Project Y mid-Sprint to a number of customers. The purpose is to do A-B testing, to crowd-source design decisions. For some features, there is no interface. The documentation is the only way the users will know the features exist.

This pre-release is due in mid-January, which is 12 working days from now. Betty has 18 days of work left. Give Alex 7 days of the work: the one hard feature (5 days) and the F1 mapping (2 days). That leaves Betty with 11 days of work, and one day to put it all together and deliver.

Alex lost 7 days of work from Project X. Between January 16 (one day after the Project Y early release) and February 15 (the day of the next Project Y release), schedule Betty to do the screenshots for Alex (returning 5 days of work). Betty can also do peer review for Alex (2 days). Betty will still have time between January 22 and February 15 to complete Project Y for the first public release.

Of course, these decisions are made with the writers and communicated to the project owners.

5.4.2. Never Say No

When you refuse to commit to a new request, it just makes the thread longer, which wastes more resources. Never refuse outright. Look at your high-level schedule and give the first open date as a possible start date. If the open date is refused, get their date. You can suggest that the requester decide which of her projects is more important and switch their schedules. If that manager does not have more projects in the existing queue, send a list of document projects that belong to other managers in the same department. You can suggest that this manager speak with peers to get a trade-off of writing resources.

"We have an important request, to create a Quick Start Guide and video for our new machine. Can we have this done by end of the quarter?"

It is April. The end of the quarter is the end of June.

"The Quick Start Guide you are talking about is full of graphics, and it takes about an hour of work to make a minute of video. I estimate this project will take: five days to write the ten top tasks, two days to make good screenshots, two days to make two or three network graphics, one day to make the video, and five days to review and implement changes. That means, this project will take three working weeks, after you get us a machine to use. Looking at our schedule, our first opening of three consecutive weeks is in August."

"Are you kidding me?? I want this in in June, and you are saying you can't start until August! That is unacceptable."

"Well, there are alternatives. I can give you a list of committed projects in your area, and you can discuss trade-offs with the other managers. We can try to schedule this project in broken days, for one-day trade-offs that do not need discussions. But if we do that, we'll need another week to make up for multitasking issues. Or you can use my schedule to escalate and request a temp for a month."

"If I get a temp dedicated to my project, can it be done now?"

"If you get budget to outsource and manage this doc project yourself, you can get it done as fast as you want. If you get approval for a temp on my team, and I have to test and interview applicants, we can probably start in forty days (ten days to find the temp, plus 30 days' notice to their current employer). If you get a machine for the temp to use. That would be closer to meeting your end of quarter deadline."

5.4.3. Real-Life Schedule Example

The previous conversation was a real-life conversation.

There was no way I was going to commit to a deadline that depended on someone else putting in a request for headcount, getting approval on time, getting me resources to find temp applicants, and pulling a prototype out of his nether regions. Let us not forget it also depended on a skilled unknown being able to start immediately.

The result of this real-life example? The headcount request was approved after two weeks, and a temp was hired four weeks after that. The machine arrived after eight weeks (two weeks after the hire, one week after the temp's first day). By the middle of June, we were ready to start. By the middle of July, the results were still not approved. Communication between the project manager and the customer had broken down for vacations and no one agreed on the final output. By the middle of August, the temp's contract was up. And here I am, with an opening on my schedule. The project began again, in August. It was completed on my first estimated due date.

The lesson learned from all the waste of time and money: "Trust Rochelle's schedule."

5.4.4. Letting Reality Play

With a granular project schedule based on the reality of known PTO, we set up our team for success.

Creating a Granular Schedule

Writers create plans for documentation projects after you confirm the high-level schedule. It is important that every technical writer uses the same tool and template. You do not want people to waste time, searching for information, when they have to do an emergency pick-up from someone else. The project plan includes a break-down of the project to smaller tasks or an outline.

In this section, we use the tasks as features and set a schedule for the project. To get the schedule in the plan properly, choose a spreadsheet tool. The instructions I give here are for Google Sheets.

In this case, we use one large project plan schedule for the whole team. Each writer has a wiki page with more notes, gathered information, pathnames for media sources, and so on. The link to the wiki page is in the Notes on the project plan schedule.

Instructions for the project plan:

1. Enter the documentation deliverables and prioritize them.
    If the printable book is necessary to sell hardware, that must get a higher priority than the PDF or HTML. It must be sent for print weeks ahead of time.

2. Enter the outline or the features for the product and for the documentation (such as graphics and glossary).

3. Prioritize product features and documentation features.

4. Next to each feature in the outline, add estimated effort.
    I suggest that the effort unit be days, and always whole numbers. The whole-number days make it easier to use the effort in formulas. If the feature is done (from a previous version), the effort is 0. If the feature will take less than a full day, join similar features of the same priority. If you cannot estimate the effort easily, give each feature one day. Your writers can give better estimates if you tell them you expect it done in one day. You will update the plan as you go.
    Plan a little, gather data, and plan again. If you wait for the perfect plan, you waste time.

5. Add the documentation features that take time: proofreading, peer reviews, and delivery creation.

6. Next to the effort, enter a status.

I suggest that you try to keep the status to three simple words. For example, start with Planned, InProgress, and Done. Or you can change Planned to Assigned, to add a motivator to assign tasks quickly.

7. Add the writer assigned to each feature.
Best: to make the next part work, use writer initials or usernames, and set the column for data validation to make sure you do not misspell a name.
Make goals for the writers of this project, in terms of improvement over the last project. Discuss lessons learned from the previous project they worked on and from the previous project of similar scope or product.

8. Add a column next to **Assigned**. Call is **Assn#** and leave it empty for now. We need it for the Start and Due date formulas.

9. Organize your people. Set responsibilities and make sure they all know the vision, the drivers of the project, the constraints, and the success criteria.

Task	Start Date	Effort (wd)	Assigned	Assn#	Due Date	Priority
Task or Milestone	Formula	Integer - work days	Writer initials	Formula	Formula	Integer

We will add more columns: Status, Notes, Contact, and you can add more as you require. For now, these are the columns we need to automate the schedule of the plan.

You can see that you will enter the task. You assign the writer, the estimated workdays, and the priority. The cool thing is that you can change any of these. The Start and Due dates update automatically to fit the schedule of the assigned writer. The Assn# column is a simple formula that gives us the dates from the assigned writer's last task. We will get into this soon.

Automating Time-Off

It is a management failure - your failure - if you must call an employee in from planned time off. Your plan includes your employees' needs. Happy, balanced employees can perform to optimal productivity. Make time-off part of your plan.

We are going to use a formula that takes into account holidays and vacation days, to make the full schedule. A realistic scope and schedule will be your starting point for the plan.

First, we're going to work out the days that employees will not work. For a large team, diverse in religion or geographic location, you do not want to type the dates in manually. We will use spreadsheet formulas to get a list of holidays first. This is always best because we all tend to forget some of them.

In the spreadsheet you started for the project schedule, add a tab or sheet named **TimeOff**. With this procedure, we will import the holidays with the spreadsheet functionality.

1. Find websites that list the national holidays of the countries and religions of the employees. Make sure the data is in a table.
   For example, if you have employees who take Muslim holidays off, you can use https://en.wikipedia.org/wiki/Muslim_holidays.

 The simpler http://www.timebie.com site is no help. It is not formatted in a table or list that works in the spreadsheet.

2. In Excel, you can use **Data > Other Sources > Web** and select the table with the data. If you use Google Sheets, it works better with the **ImportHTML** function. In a working worksheet, in the first cell of data, enter this formula:

```
=IMPORTHTML("URL","table")
```

For example, to get the USA Holidays of 2018:

```
=IMPORTHTML("http://www.officeholidays.com/countries/usa/
2018.php","table")
```

3. When the data loads, you cannot delete the columns you do not want.
   In Excel or Google Sheets, you can copy the date and holiday name columns and then paste the values (**Paste Special > Values Only**) in the TimeOff worksheet.
   In Google Sheets, you can load the data to a range and then use the QUERY function to pull only what you want. For example, in column J, enter:

```
=IMPORTHTML("https://en.wikipedia.org/wiki/
Muslim_holidays","table")
```

The result is something like this:

	Hijri date	1442 AH
Islamic New Year	1 Muḥarram	20 Aug. 2020
Ashura	10 Muḥarram	29 Aug. 2020
Arba'een[a]	20 Ṣafar[b]	8 Oct. 2020
Eid-e-Shuja'[c] (Eid-e-Zahra)	9 Rabī' al-Awwal	26 Oct. 2020
Mawlid an-Nabī - Sunni	12 Rabī' al-Awwal	29 Oct. 2020
Mawlid an-Nabī - Shia	17 Rabī' al-Awwal	3 Nov. 2020

Birthday of 'Alī ibn Abī Ṭālib[a]	13 Rajab	25 Feb. 2021
Laylat al-Mi'raj	27 Rajab[e]	11 Mar. 2021
Laylat al-Bara'at	15 Sha'bān	28 Mar. 2021
Birthday of Muhammad al-Mahdī[c]	15 Sha'bān	28 Mar. 2021
First day of Ramaḍān	1 Ramaḍān	13 Apr. 2021
Laylat al-Qadr	19, 21, 23, 25, 27, or 29 Ramaḍān[f]	1 - 11 May 2021
Chaand Raat[g]	29 or 30 Ramaḍān[h]	12 May 2021
Eid al-Fitr	1 Shawwāl	13 May 2021
Hajj	8-13 Dhū al-Ḥijja	18–23 July 2021
Day of Arafah	9 Dhū al-Ḥijja	19 July 2021
Eid al-Adha	*10 Dhū al-Ḥijja	*20 July 2021*
Eid al-Ghadeer[a]	18 Dhū al-Ḥijja	28 July 2021
Eid al-Mubahalah[a]	24 Dhū al-Ḥijja	3 Aug. 2021

4. The results show the Hijri date, which you cannot use in this solution. It also shows the dates for 4 years. Right now, you only need one. Let's filter the dates. In column A, enter:

```
=QUERY(J1:N20,"select J, L" , 0)
```

The results show only the name of the holiday and the Gregorian date.

The Query formula makes it easier to paste the values of dates when they are not in adjacent rows. But in the end, we have to have only the date values, not the results of formulas. Each employee will enter their own time off (such as doctor appointments and non-holiday vacations). The dates must be in chronological order for the whole solution to work. You cannot sort a mix of automated output and manual input.

To personalize the time off:

1. Get the dates for the national, religious, and cultural holidays that are relevant for your people.
    Send the holidays to your employees, with a column for each employee. Ask them to mark which holiday they will not work. In many scenarios, you will have to keep this

data private. Send a copy of the sheet to each employee, or do this step with each employee at your desk, behind closed doors.

For example, for a team in Israel, collect the Israeli national holidays, Jewish holidays, Islamic holidays, and Christian holidays (if all three are represented on the team). Some holidays are more than one day. The company may give Paid Time Off (PTO) for a half-day of the date of the first day. Some employees will decide to come in for the half-day, and others will take the half-day off (not come in at all). The company may give PTO only for the first and last days of a week-long holiday. Your employees can use this to tell you if they intend to take the complete week off with their vacation days.

2. On the TimeOff sheet, each employee has a column with their name or initials as the header.

3. Each employee enters (or copies-pastes) the dates they will be off, in the cells of their own column. They must enter only the dates, not the reason. And the dates must be all in the same format.

4. Sort each column by earliest to latest.

5. Add a row above their names, before the first date. Enter the value for the weekends:
    Common Weekends

    - 1 = Saturday and Sunday (default)
    - 7 = Friday and Saturday (Israel and Jordan)
    - 11 = Sunday only (India)
    - 17 = Saturday only

    Less Common Weekends:

    - 2 = Sunday and Monday
    - 3 = Monday and Tuesday
    - 4 = Tuesday and Wednesday
    - 5 = Wednesday and Thursday
    - 6 = Thursday and Friday
    - 12 = only Monday ... 16 = only Friday

    If you have a part-time employee, you can use the seven 0/1 method of Google Sheets. This is not supported in Excel 2010.

    Get the full list of valid values for weekends from the Microsoft Office guide or from the Google Guide.

    Here is the TimeOff sheet for Alex, Betty, Charlie, and Deepak. The first row under their initials is the number that represents their weekends. Alex's weekend is Friday and Saturday. Betty and Charlie have Saturday and Sunday off. And Deepak has only Sunday off.

7	1	1	11
AZ	BY	CW	DU
30 Jan 2019	1 Jan 2019	1 Jan 2019	3 Jun 2019
7 Feb 2019	21 Jan 2019	13 Mar 2019	4 Jun 2019
21 Mar 2019	18 Feb 2019	4 Jul 2019	11 Aug 2019
16 Apr 2019	16 Apr 2019	28 Nov 2019	19 Aug 2019
18 Apr 2019	12 May 2019	25 Dec 2019	25 Aug 2019
21 Apr 2019	27 May 2019		
22 Apr 2019	4 Jul 2019		
23 Apr 2019	14 Oct 2019		
24 Apr 2019	28 Nov 2019		
25 Apr 2019	29 Nov 2019		
8 May 2019	25 Dec 2019		
9 May 2019			

Important! The dates must be in order. If a writer enters new Time Off dates out of order, the schedule will show an error. Our schedule depends on a named range of dates for each writer. When you have the range, you can sort it easily.

To save the time off as objects for formulas:

1. Create a named range for each employee's column of days off.

2. Name each employee's range of dates as: *name*_TO
   The name here must match the name we will use in the rest of the solution. If you use full names, use an underscore, or remove the space. I usually use initials to make it easier.

   - To name a range in Google Sheets, click **Data > Named Ranges**. In the panel that opens, enter a name, and make sure the range includes the cells with only the dates.

   - To name a range in Excel, select the cells and then enter a name in the top left corner of the sheet.

3. Name each employee's weekend cell as a range: *name*_wkend

In Google Sheets, I added these instructions to my team, on the TimeOff sheet:

```
How to fill in your days off:
1. Add the new date to your column.
2. Click Data > Named Ranges.
3. In the list, click your range. Do not change the name! It is
important.
4. Click Data > Sort Range.
```

We are ready now to create our schedule formulas.

Automating Schedules

The next steps are where the magic starts. It is going to get complicated. Stick with me. After this is done, your team schedule will be accurate. While other managers give ETAs such as *the middle of March*, you will be able to say *March 13* with the conviction of automated planning.

To fill in your team schedule:

1. In the main tab of your Project Plan Schedule, where you have the column headers already, under **Task**, enter **Start**. In the next cell (Start column), enter the start date of the plan. The **Priority** cell is 0 (zero). The rest of the row is empty.

    To get the default date format, you can use =today(), copy the value and paste **Value Only**.

    You can use today() to set the start date. But if you do, you or your writers must update the effort as you complete parts of tasks and remove rows when tasks are completed.

    I like to color the cell, to show that it is a manual date and meant to be updated by hand if necessary.

Task	Start Date
start	16 Dec 2019

    I keep a color legend at the top of the sheet.

    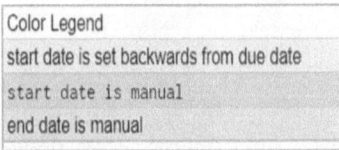

Color Legend
start date is set backwards from due date
start date is manual
end date is manual

    Then a few empty rows and then the headings. Thus, the first actual task (and date formulas) is on row 10.

2. Set the Start date as a Named Range called **DefaultStart**.

3. In the first column, **Task**, your writers fill in the tasks or outline of their projects.

If the tasks are broken down to things that will take less than a day, work with the writers to combine tasks to make at least one day. A task can take more than one day, but the formulas will be broken if they are not at least one day.

4.   Tell the writers to skip the dates. In the third column, they enter the estimated effort for the task.

     If they do not know, look at the TimeTracker. Find a similar task and get the average time it took. If there are many to choose from, the min and max or median. If there are none, start a discussion with 1 day. Remember, this is accumulated hours for one project. It cannot include time waiting for SMEs or buffers for risks.

5.   In the **Assigned** column, they enter their initials. Make sure the next column is not touched!

     Their initials (or names, if that is what you use) must match exactly the way you used them in the Named Ranges of the TimeOut sheet.

6.   Let's start with the easiest part of the date formulas: tasks that start on the next available day.

     (Before we are done, we will also explain how to set the start date backward from tasks that have a hard-stop due date.)

     In the **Start** column of the first task, with the row of 10, enter:

```
=workday.intl(iferror(index(D
$9:F10,match(concat(D10,countif(D$9:D10,D10)-1),E
$9:E10,0),3),DefaultStart),1,indirect(textjoin(,false,D10,"_
wkend")),indirect(textjoin(,false,D10,"_TO")))
```

To understand this, let's go from inside out. The workday.intl syntax is:

```
=WORKDAY.INTL(start_date,days,[weekend],[holidays])
```

Let's look at the weekend first:

```
indirect(textjoin(,false,D10,"_wkend"))
```

We use textjoin to get the weekend of each writer. The textjoin syntax is:

```
TEXTJOIN(delimiter, ignore_empty, text1, [text2, ...])
```

In our usage, we do not have a delimiter. We enter a comma (no space). We do not want to ignore empty cells. We enter `false`. The writer initials are in column D. We enter `D10` (we are on row 10). Do not use the $ sign. Then we join the writer with "_wkend" to get the *writer*_**wkend** name.

The textjoin formula is wrapped in the **indirect** formula. This turns the string of *writer*_**wkend** to the value of the range.

The result is the value of the weekend for the assigned writer. If the writer is Alex (AZ), whose weekends are Friday and Saturday, the value is **7**. You will not see this. But Alex will never be assigned a start date or due date on a Friday or Saturday. If you change the assignment to Betty, the weekend automatically updates to Saturday and Sunday (1).

Let's look at the holidays:

```
indirect(textjoin(,false,D10,"_TO"))
```

Well! Look at that! It's the same as the weekends, but it calls the **TO** ranges rather than the **wkend** ranges. Easy peasy.

Let's do another easy one - the number of days before the task starts.

```
=workday.intl(start_date,1,weekend,holidays)
```

We add 1 day to the start date. Our start date will be the date that the writer completed the last assignment. We let the writer complete a task and start the next one on the next workday.

Now for the hard part: setting the start date according to the next available date of the assigned writer.

a. In the column next to **Assigned**, enter this formula:

```
=concat(D10,countif(D$9:D10,D10))
```

This means: Count the assignments this writer has from D9 (where we started with the Start date, which is empty) to this cell. Notice that we must use the $ sign this time. In the weekend and holiday formulas, we wanted to use the same row. This time, we want all the rows to start the count from D9. The dollar sign is required.

The countif formula is the second part of the concat formula. The result is the writer's initials (D10) and the number of assignments the writer has in the plan. For example, if Alex has the first assignment, D10 = **AZ** and D11 = **AZ1**.

When we copy this formula down, the count range is updated. For example, if the last task is on row 15, the formula is automatically (on paste) updated to: `=concat(D15,countif(D$9:D15,D15))`

b. In the inner part of the start_date of the workday.intl formula, we see this repeated:

```
countif(D$9:D10,D10)-1
```

This means: Get the count of assignments (including this one) and subtract one.

c. Now we are going to use index and match. This is the best way to get values from around a spreadsheet, to use in formulas. Match finds the cell that matches the data we are looking for. Index gives the value of the cell.

The match syntax is:

```
MATCH(search_key, range, [search_type])
```

We use:

```
match(writerNum-1),E$9:E10,0)
```

We know the inner formula equals the writer and number of assignments minus 1. If this is the first assignment, it will equal *writer*0. That is OK. We get into that in the next steps. If Alex's second assignment is on row 16 (E16 = AZ2), match will search for AZ1, in the range of E9 to this row. On row 10, this is E$9:E10. On row 16, this is E$9:E16.

The 0 (zero) at the end is required. It means that the search range will not be sorted. We will sort by Priority, not by writer. If you forget this zero, nothing will work.

Now we have moved our formula to look back and find the row of the last assignment of this writer. A search for AZ1 will result in match = E10.

The index syntax is:

```
INDEX(reference, [row], [column])
```

We use:

```
index(D$9:F10,match(row of last assignment of this
writer),3)
```

This means: In the last assignment (Column E) of the writer (Column D), give us the date they ended that assignment (Column F). Index looks at D, E, and F, and returns the third value of the matching range.

I learned how to do this backward time calculation from Ben Collins. I don't remember if it was a course or his free newsletter of Google Sheets tips. But I will always remember he is *The Best Resource* for spreadsheet knowledge and training.

d. If this is the first assignment of the writer, match will look for *writer*0 and it will be an error. To fix this, we wrap index and match in iferror, with instructions to use the DefaultStart date if it is an error:

```
iferror(index(date of last assignment),DefaultStart)
```

Whew! Well, the hard part is done. Now you have it, and you understand it. Let's continue.

7. In the cell of the **Due** column, we will use the workday.intl formula as it was meant to be used. Enter it now:

```
=WORKDAY.INTL(B10,C10,indirect(textjoin(,false,D10,"_wkend")
),indirect(textjoin(,false,D10,"_TO")))
```

B10 is the start date that we figured out in the previous step. C10 is the effort that the writers entered. Then we have the weekend and holidays as before.

8. Copy the Start date down. You can double-click the crosshair or drag it down, or copy the cell and paste it. Copy the Due date down. Make sure the results make sense and fix any errors.

Your basic automated schedule is ready. If you change the assigned writer, the schedule updates automatically for the new writer. If deadlines are tight, you can quickly juggle assignments between writers before you ask for trade-offs from other teams. And now you and your team know exactly what is expected of them.

Automating Hard-Stops

Some projects, especially shorter projects, such as release notes for sprints, have committed due dates. For these, we will manually enter the date in the Due column. We will use a formula to show us on which date the work must begin to complete it on time.

To set the start date backward from the due date:

1. In the row where the task is given, enter the due date manually.
   Make sure it is in the same format as the automated dates.

2. Enter the effort and writer initials. The *writerNum* cell fills automatically.

3. In the **Start** column, enter this formula (replace 19 for the row number you are on):

```
=workday.intl(F19,-
C19,indirect(textjoin(,false,D19,"_wkend")),indirect(textjoi
n(,false,D19,"_TO")))
```

We see the weekend and holidays formulas that we learned in the previous procedure. That leaves:

```
=workday.intl(F19,-C19,weekend,holidays)
```

F is the column that holds the Due Date. C is the column that holds the effort, in the number of workdays. This means: Show us the date that is the due date minus the days required to complete the task.

If the due date is Wednesday, August 7, 2019, with an effort of 2 days, without weekends or holidays in that week, the result will be Monday, August 5.

Task	Start	Effort (wd	Assigned		Due
Z RN	5 Aug 2019	2 DU	DU3		7 Aug 2019

It is important to mark the manual date, to set it off from the automatic dates. If you do not, it will be lost when someone (invariably) pastes the formulas again.

Automatically Sorting by Priority

The schedule depends on the order of the tasks. Start dates are one day after the due date of the previous assignment. But hard-stop tasks, with committed due dates, can fall anywhere. You do not want to sort the tasks alphabetically, even if you change the task description to include numbers or sort indexes. That causes a lot of manual work. You do not want to sort by writer initials, which would give you a shift team. (Your writers can save filters for their own assignments to see their schedules, without affecting others.) And you do not want to rely on tasks coming in the correct order, which they never do.

We are going to add a column for Priority and sort by it, with a Google Sheets macro.

The value of the **Priority** column is an integer.

- Zero is reserved for the start date. We want DefaultStart to stay in place, at the top.

- Tasks that we know of, but which are not ready for documentation, we use 99 or 999, depending on the size of the team and the number of tasks.

- Tasks that are started or which must be completed immediately, we use an integer between 1 and 5. I like to make sure each writer does not have more than one task of the same 1 - 5 priority, to help them decide what to do next. But it is not a hard-fast rule.

- Tasks that belong to one release milestone have a common digit. For example, all the tasks for release Product A version 2.1 EA release have a priority between 51 and 59.

- I reserve 50 and 60 for milestones. For example, the release of 2.1 EA is set to 60, with a manual due date and zero effort. If tasks are added to this release, I will see immediately if we are overcommitted. if tasks of priority 59 have a calculated due date that is after the release, I will juggle assignments. If that does not work, I will discuss an impact plan with my manager.

An impact plan means that some features are moved to a secondary phase of delivery or dropped from the project. Before a manager agrees to this, they usually want to change the priority of tasks. To make sure they intuitively understand the priority meanings, we add a column for **Release**. Sometimes our manager, or the project owners, already decided that a feature will not be in this release. A feature can be important but deferred. To help others set the documentation priorities, we show them what we know of the features for each release.

Task	Start	Effort (wd	Assigned		Due	status	Priority	Release
start	Jun 13, 2019						0	
ABC	16 Jun 2019	1 AZ		AZ1	17 Jun 2019	Working	1	flagship
D Feat	14 Jun 2019	1 BY		BY1	16 Jun 2019	Working	4	flagship
E Feat	14 Jun 2019	2 CW		CW1	17 Jun 2019	Working	4	flagship
F Feat	14 Jun 2019	5 DU		DU1	20 Jun 2019	Working	4	flagship
Y - A	21 Jun 2019	1 DU		DU2	23 Jun 2019	Planned	100	Y
Y - RN	17 Jun 2019	1 BY		BY2	18 Jun 2019	Planned	201	Y
Y - videos	18 Jun 2019	1 AZ		AZ2	19 Jun 2019	Planned	203	Y
X	20 Jun 2019	1 AZ		AZ3	23 Jun 2019	Planned	203	X
Z	18 Jun 2019	1 CW		CW2	19 Jun 2019	Planned	205	Z
Z RN	5 Aug 2019	2 DU		DU3	7 Aug 2019	Planned	210	Z

If the company decides that Product Z is to be released early, before Product Y, update the Priority and then sort the table. We do this so often, that I automated the sort.

To automate the sort by priority in Google Sheets:

1. Enter this script in the Google Sheet:

```
function SortTasks() {
  var spreadsheet = SpreadsheetApp.getActive();
  spreadsheet.getRange('tasks').activate();uo
  spreadsheet.getActiveRange().offset(1, 0,
spreadsheet.getActiveRange().getNumRows() -
1).sort([{column: 8, ascending: true}, {column: 9,
ascending: true}]);
};
```

2. Select the table, including the column headers, and save it as a Named Range.

3. Create a button that runs the script.

4. Allow access (you, and each writer, do this step only once).

To run the automatic sort:

1. Change the priorities.

2. Click Data > Named Ranges and make sure the range for the table includes all newly added tasks.

3. Click the sort button you created.

Done and done! You, your writers, and your managers can easily update and sort priorities, to get an updated schedule. Instead of dealing with dependencies and inner properties of a Gantt chart tool, you change priorities to show when a specific order is necessary.

5.4.5. Communicating Overtime

You will probably notice that most of your writers are older than their SMEs. Hi-tech is still a young person's game. Developers, testers, even CEOs get started straight out of high school or military service. You will find many are working as they study for their first degree.

The myth of tech writers around the world is they are in a different stage in their lives and careers, and so they cannot be expected to work like the hungry youth.

Nonsense.

Many of the "young" people you see in corporations have families and dependents. Most get just as frustrated as writers when their work depends on others.

The difference is if QA or R&D employees do not put in the time and effort to complete their work, the product or its timeline suffers. If writers do not put in the time or effort, and if no one pushes for review, the users and Support suffer after the release. If someone notices that the documentation failed, it is usually too late to resolve for that version.

How do we as managers talk to our writers when they use overtime inefficiently, or refuse to put in the extra effort when it is needed?

1. Get the stats from HR.
    What percentage of the QA and developers have children? Get that answer. It is the best way to stop the incorrect thinking that the writers are more pressured by life.

2. Talk about the product timeline.
    There are tasks that must be done, even when there is no new release to document. Then there is a peak time when the release date is driving everyone.
    When the developers are creating the product, writers develop their plans, standards, knowledge, and methods. Before the first QA drop, writers have an approved outline. When the product goes for testing, the peak time starts. This is when you expect 100% effort during working hours.
    You want the project done on time. View forced overtime as a planning failure. If writers are working overtime during non-peak periods, find out why. Some people

simply like to work, but most will burn out if they work too long, too many days in a row. Make sure overtime is saved for peak times and for necessary intensive fixes.

Prepare your team for the peak time coming: "Our daily tasks will be bigger. We have to cut up the final desired results into days. And we have to do the daily tasks until they are done. If you complete your part in the regular working hours, that's perfect. If you do not, put in the time you need to complete them. Keep notes on what went wrong. We will use them to fine-tune our planning needs for the next time."

When you schedule a project, identify resource needs. Make a table of all the features and the number of estimated days of effort for each feature. Divide the sum by the number of writers.

3.   Prepare for negative reactions.

Do not get emotional or angry when someone refuses. Your answer can be this: "I know you want to succeed in your work. I see the effort you make, so I know you want the company to succeed. That's why your reaction surprised and disappointed me. I expect more. And I know: you can do this."

Use "disappointment" as an opening to resolve the conflict between what you expect and what they are willing to give.

"This is what we need to succeed as a team."

4.   Define success for your team as being part of the whole, doing your part, to the best of your ability. You all want complete documentation for the agreed-on feature list, tested for accuracy, consistency, and proofread.

Remind your employees about why the company deserves their commitment - its product, its internal core values, its value to society, whatever makes your employer worth your loyalty.

Show them the competition in the industry, to challenge their preconceptions of what is good enough and push for their own higher expectations.

Open a discussion about how the documentation can raise the level of the product.

Motivate your team with project engagement.

If your team cannot make the deadline, prepare alternatives (deliver in stages, drop lower priority features, or bring in more writers). Bring that up immediately with the project owner.

If the owner agrees to a prioritized list of features for the first phase, make sure the writers know it and do not waste precious time on unlisted features.

Some project owners jump directly to the hire of a temporary contractor. It might be an option, but you and the project owner must factor in the time it takes to train people in your tools and the company technology.

Make sure everyone on the team knows their tasks and the life-cycle of their parts. For example, if two writers will update all the documents for a new user role, they each must know the division of documents or features and track the status of each. They will double the research because each must learn it. Then they share the writing, testing, proofreading. They each track documents sent for review and when a document's changes are implemented.

The hours you expect from your people are based on the needs of the product and the plan to reach the first draft date. If someone falls behind in the schedule, they can expect to work overtime. At this point, you and the writer can take an hour to understand why they fell behind. After you understand what happened, make a decision: Does it affect the rest of the schedule? Do you need more people on this project? Should you juggle writers? All of this is up to you. Do not bring in the writers until you make a decision and need their actions or change of plans.

5.4.6. Follow Up

This is the difficult part for innovation-driven managers. We want to spend our days on new and exciting things. But you must follow up. Imagine that you set out detailed plans and the perfect schedule, and the next week, you have no idea what to say when your manager asks for status. Embarrassing!

Follow up on a routine basis. Writers have a schedule and a plan with every task to do, time allocated to do it, and a date to have it done. Experienced writers can make this for themselves. For writers new to you, lead that plan. Make sure you have easy access to the plans and schedules. If the project is one work-week, meet casually for five minutes every day. Maybe just pop in with "Are we on track? Do you need help with something?" If the project is months long, meet once a week.

Be careful with the follow up! Control of a project and its sub-tasks is essential for quality output. During these routine follow ups, let the writer talk. Make no judgments. Ask questions for more information. If you think the writer is not focused on the priority issues, ask why they believe the task at hand is more important.

I suggest that these follow ups not be by email. That stinks of micromanagement. If you can't meet face-to-face, use an instant messenger or meeting software.

5.5. Working the Plan

Make sure the writers agree with your estimates. If the plan exceeds the product deadline, ask the writer if some features can be done in half a day or less. If the writer says some features will take more than a day, enter that in the plan. If the writer is uncomfortable with day estimates, you can try for hours.

5.5.1. Assess Required Writers

Use a formula to calculate how many writers you need to meet the deadline. Calculate how many working days you have, from the day after the team meeting, to the deadline of the product. You know the effort (required days to complete the work). Now you can see how many writers you need. For example, if you have 20 working days before your pre-GA deadline, and 320 estimated hours of effort, you need 2 writers (320 hrs / 8 hrs per day = 40 days effort / 20 days available = 2 writers).

5.5.2. Get Engagement

Review the plan, face-to-face, with the people working on it. If you have a remote meeting, use screen sharing. Everyone will see your cursor moving and the different sheets and tables in focus while you explain the plan.

It is important that everyone on your team uses the same plan template. If your team is keeping a plan for each project in one central place, they can pick up from each other and return to a dropped project with little or no involvement on your part.

Alex is working on X. Betty is working on Y. Alex goes on a planned vacation. When he comes back, he skims his project plan and gets back to work with minimal downtime. Betty is asked to present in a conference in Las Vegas. While she is gone, Charlie picks up her projects. Charlie can look at Betty's plan and know what to do, quickly getting the latest versions of the correct source files.

The plan must change during the project for changes in the product and for updates in progress. When a small change is made, make sure everyone can see the change in the plan. When a significant change is made, hold a face-to-face meeting. Get a consensus, in writing, before you begin.

The plan includes when you will send topics for review to SMEs. Make sure the writers share the plan with their SMEs and get commitments for review. If an SME does not deliver on the committed deadline, the writer sends a polite reminder. If the writer does not get a response, send a terse message: "When can we expect x?" The writer brings the response (or no response) to you. Whether the issue should be escalated or not depends on the priority of the project, the confidence level of the writer in the accuracy of the documentation, and the personalities of the writer, the SME, and the manager of the SME.

The only advice I can offer is: use escalation carefully. Consider the feelings and responsibilities of everyone involved.

5.5.3. Peer Review

Encourage, even enforce, peer review. We catch other people's typos that we miss in our own writing. We learn from each other. We become better writers by becoming better readers. Do not let your team put this off! Of course, we are busy. We have schedules and plans. But peer review must be done. If not, we lose out on the quality of the current projects and on improved quality writing of future projects. Consider setting a number of hours for each week to work on peer review in the overall Expectations, and make sure that time is in each writer's weekly schedule.

Tip: Use a system to set off types of comments. For example, comments for errors start with 1. Comments for technical questions start with 2. Comments for style start with 3.

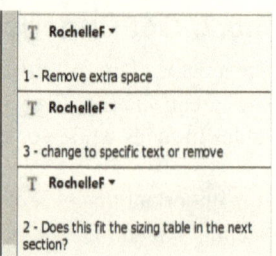

```
The value of ulimit is too low for the Agent to run processes that monitor and analyze the behavior
of other processes. ulimit must be less than file-max. The default settings are often not high
enough for high-performance servers.

Increase ulimit to 64 times the number of CPUs. Make sure file-max is greater is than ulimit.

# sysctl -a | grep fs.file-max
value of file-max
1.   Increase file-max with these commands, where 500000 is an example of a high value.

     # vi /etc/sysctl.conf
     fs.file-max = 500000
     # sysctl -p
```

T RochelleF ▾
1 - Remove extra space
T RochelleF ▾
3 - change to specific text or remove
T RochelleF ▾
2 - Does this fit the sizing table in the next section?

5.5.4. Documentation Project Plan Template

The Plan can be a wiki page, a document, or a spreadsheet. Whatever you choose, make sure everyone uses the same template.

Project description. Is this the next version of a release? What is the main purpose of this project? What is the primary driver? If you have the filled document request form, you can paste it in the Project Plan.

Release Manager	*Name, contact link*
R&D Owner	*Name, contact link*
QA Owner	*Name, contact link*
Writers	*Names*
Target	*ClearQuest, JIRA, or similar - Target name and link to see all features and issues*
Queries	*ClearQuest, JIRA, or similar - links to queries relevant to writers (Assigned to Team, TechPub project issues, RN issues, and so on)*
Task Schedule	*link to daily schedule to complete the project*

Deliverable Status Table

MS	Name	Upload	Src	Pub Profile	1st	2nd	Final	Due	Notes
EA GA	*file*	*link*	*ID*	*list*	*Status*	*Status*	*Status*	*Date*	|

Documentation Workflow - *Deliverable Milestone*

☐ Commit to request

☐ Kickoff meeting

☐ Make queries and link to plan

☐ Get product or prototype

☐   Write and proofread features

☐   Sent features for review: CL-checker tool, peer review, SME review

☐   3 Drafts

☐   Upload and commit to repository

☐   Clean up: delete temp topics and VMs, update plan for notes and status, write lessons learned

Notes

* The tables and lists are clean, without notes.
* Writers keep notes in a separate section.

Resources

* Links to other documents or videos
* If the plan is on a wiki, add a macro to show attachments

Notes for the High-Level Status Table:

MS is the release **milestone**. Our deliverables often change between Beta, Limited EA, EA, GA, Service Pack, and so on. You must keep a repository (wiki attachments, Google Drive, shared folder, or similar) of each deliverable released.

Upload becomes a live link when the documentation is delivered.

Src is for CCMS tools (AuthorIt, Paligo, or similar). Each object has an ID. This is the ID and link to the container object. If you use a document-based tool, this is the link to the document.

Pub Profile is the conditional text (profiling attributes, variants, or similar) and format (PDF, HTML, or similar, and the template for each) to select to publish this deliverable.

The draft reviews are labeled according to your company and team policy. Here we used **1st, 2nd**, and **Final**. The **status** is from a list that all writers use: *Planned, InProgress, SentForReview, Comments* (the writer implements changes from comments received), *Done, OnHold* (the reviewer did not return comments on time).

Due is the date you committed to delivering the documentation. It is not the same as the release date.

5.5.5. Create a Risk Management Plan

Writers with less management experience often put the risks inside the plan. They add buffers for possible illnesses or network issues. Do not do this.

Create a Risk Management Plan. Do this with the writers. They need to know that their experience is considered.

> I like to keep this with the project plan, but traditional PMs say it should be a separate document. Up to you.

For each known risk, guess how likely it is to happen. Estimate how many days impact it will have on the plan. Then brainstorm to come up with solutions.

For example, if a project is planned for an update to a version that is in development, the risk is that the first product will require more work after the start date of the update. You asked around (discreetly) and find that this is very likely. You look at earlier plans of similar projects, and you see there is an average of 6 days of work after a release date. You and your team think of different solutions. This is the Risk Management Plan for Update Y to Version X:

Risk	Likelihood (1-5)	Impact (days)	Mitigation
VerX not done	5	6	Complete multiple tasks in one day
			Drop lower priory tasks to post-release publish
			CR Monday to fix released docs
Unplanned time off	3	3	Have second writers to handle immediate issues
			Get VPN to work from home

We can actually put some of the mitigation into action before the risk happens. We can ask for the writers to get VPN access. If a writer must stay home with a sick child, they can work from home on task types that we set as possible for remote work.

5.5.6. Sprinting is a Solution

With Agile models becoming more and more popular, "sprint" might not be the best word here, but it's the word I know.

In a documentation sprint, the team works together, intensively, to complete a priority project. The doc sprint is a workaround to a breakdown in plans. It is a possible mitigation for a large-impact risk that is close to the delivery due date.

For example: An important potential customer wants to run a PoC. The first documentation milestone is a month later. You run a sprint to meet the new demand.

To make a sprint work:

1. Decide on one goal for the team sprint.

    - It must have a higher priority than all the other current projects. Don't let a loud manager push you into a sprint. Make sure you agree on the priorities.

    - It must have a low enough difficulty for all writers to handle. If some parts of the project are difficult, see if the other writers can provide real added value on the easy parts. It's no help if one writer wastes time teaching the technology to others.

2. Design communication.
    How will all writers work simultaneously on one project, without overwriting each other or duplicating efforts? Identify parts that are dependent on the completion of others. Can you assign all the parts of a linear flow to one writer? If not, can you assign them to writers who work in shift-like hours.

3. Set a specific start time and end time for the sprint, but don't tell the team yet.
    One cool thing about sprints is they can be a bonding experience. But there is always at least one person who wants to get a head start on everyone else. Discourage this. Remind them of their own planned priorities.

4. Add the sprint time to the master schedule.
    Do not overwrite the scheduled commitments. Push them up. If you will have a delay on a commitment because of a sprint, discuss with the managers first. If there is a

delay that is one day or less, there is no need to communicate. A trade-off discussion for one day is a waste of time. Estimate an hour for each person in the sprint to get back to the scheduled plans after the sprint. If there are 8 writers in the sprint, that is one calendar work-day to change focus.

5. When you have an approved schedule and plan for the sprint, notify the team.

6. Set one employee, usually the owner of the project, to be the point of contact for all questions. This person will give everyone prerequisite data and materials.

7. Make sure all the writers track their time.
   Everyone escalates issues immediately to you. Do not let anyone stop to investigate details. Make sure everyone works through questions and missing information, sending the questions to the appropriate R&D or QA contact but not waiting for answers to continue.

8. Summarize status and issues at the end of each day.
   Make sure you are on top of the project and the sprint effort. It is too expensive to leave on its own.

5.6. Closure

At the end of the project, document issues on which to improve or for which to plan for the next project.

5.6.1. Improved Estimates

Improve your ability to estimate effort. If you can say exactly how many days or hours it will take the team or specific writer to complete the scope, you don't have to worry about fast. The project will be done on time, or you can raise the red flag from the very beginning.

5.6.2. Planning for Post Release Feedback

Define what priority feedback is and what is non-priority.

Sample Feedback Priorities:

1. Customer feedback

2. Support or Sales feedback

3. Internal bugs with highest priority

4. Internal bugs that must be fixed before next version

5. Internal feedback that is less urgent and more difficult

Set a time for the writers to complete each type of feedback. This is your team's Internal Service Level Agreement.

For example, you say that customer feedback must be investigated and applied to released documentation in less than 48 hours.

Make sure your team understands this is an internal SLA. "We have an SLA," must never go in emails or customer-facing communication. That will open your team, and your employer, to legal commitments.

5.6.3. Reach Higher

Use each project as a step to improve the next project.

• Nothing succeeds like success.[4] During a project, when a writer does something very well, or an SME tells you something good about that person, share the joy. Let your writers know that you recognize their efforts and appreciate their successes. Their work will rise in productivity and quality. Be careful to be honest, always. Do not use "show recognition" as a tactic to get more out of demoralized employees.

---

[43] Helps, Sir Arthur. *Realmah,* 1868. "Rein ne reussit comme le succes."

- When a project is completed, well and on time, show appreciation. Give recognition to the writer, in a manner that is fair and best for that person. If the SMEs or other people helped a lot, thank them in an email, with their managers in CC.
This is important for technical writers. When project managers or sponsors hold release parties or reviews, we are most often forgotten. Do not expect others to show appreciation for your team's work.

- Review the project with the writer. Let them lead a discussion about what worked and what did not. Ask them to document this. Show them all that you did in the management of the project. Let them take more control on the next project. More control over daily tasks raises morale, energy, and quality.[5]

---

[54] James Adonis claims that control leads to quality (http://www.jamesadonis.com/Newsletter_Motivation.htm), but the source he used (I think this was the source: https://pdfs.semanticscholar.org/3f44/f35d1779ea91896c9f443904aab90c2d9511.pdf Cerasoli, C. P., Nicklin, J. M., & Ford, M. T. 2014) seemed to say that more analysis was necessary to make that claim.

5.7. Power Up!

When more than one person has to do the same thing multiple times, it is time to make some macros for the team. Be smart about this: use an engineering cycle. You will not save time if you try to shortcut your own research and development.

Visual Basic for Applications (VBA) is the language that Microsoft uses for macros in Office applications of version 2007 and later. It is not hard to learn. The easiest method is to record a macro in a document and then edit the syntax that Word created automatically.

Google Apps use a form of Java Script. The easiest method is to find an App Script online that does what you need.

To create macros or scripts on a template, first sandbox the code:

1. Save the template as a copy.

2. Research what is possible and what is done.
    Look for scripts or macros online. There is no need to recreate the wheel. If you use a macro that someone else created, put their name or URL in the code. Make sure it is free to use.

3. Document what you want the macro to do, where you found information, what works, and what changes you made.

4. Test the macro on the copy.

Now you are ready to go to production:

1. Save a backup of the template file.
    Tell your team you are working on the template. (Best: do this work on the template when the team is not working, such as evening or lunch hour.)

2. Move the macro to the actual template.

3. Test again. If you make a mistake, delete the template and restore the backup.

4. Tell the team what macros you made, the steps they will automate, the benefit of using them, and how to test if they work. Ask others on the team to test it.

5. When enough of your team agrees that they work and are useful, tell them how to make buttons on their local version of Word (or Google Docs or Google Sheets) to access the macros.
    You can make a new tab, with your own ribbon, a new group in a selected ribbon, or add the buttons to the Quick Access toolbar.

To set up Quick Access buttons for Word macros:

1. Open a Word document.

2. Right-click in the Quick Access toolbar and select **Customize Quick Access toolbar**.

3. In the **Choose commands from** list, click **Macros**.

4. Double-click the macro you want. If you are not sure you have the right one because the names are too long, you can see the full name in the **Modify** step soon.

5. Select the macro in the list of selected actions and click **Modify**.

6. Go to the end of the **Display** name and make sure this is the macro you want.

7. Enter a name, or delete the prefixes, to see the macro name. (Best: choose an icon too.)

8. Click **OK**.

5.7.1. Example: Finding Matching Values

Here's an example of a power-up system. We had an ongoing translation project. Everything in English was translated to Japanese. We had an issue with the translation memory. We were using AuthorIt, which has IDs for topics (each topic has multiple sentences). Translation is by sentences. When a project included already-translated topics, we had a mismatch.

Issue: Same topic has *incorrect ID* in JP, but same GUID as in EN.

Solution: Parse XML topics for GUID, ID, and Description (for testing).

Here's the XSLT parser, for AuthorIt:

```
<?xml version='1.0' encoding='ISO-8859-1'?>
<xsl:stylesheet version="1.0" xmlns:xsl="http://www.w3.org/
1999/XSL/Transform" xmlns:xs="http://www.w3.org/2001/XMLSchema"
xmlns:ait="http://www.authorit.com/xml/authorit">
<xsl:output method="html" />
<xsl:template match="/ait:AuthorIT">

  <HTML>
   <BODY>
    <TABLE>
     <TR>
      <TD><B>GUID</B></TD>
      <TD><B>EN/JP ID</B></TD>
      <TD><B>Description</B></TD>
     </TR>
```

```
<xsl:for-each select="ait:Objects/*">
 <TR>
  <TD><xsl:value-of select="ait:Object/ait:GUID"/></TD>
  <TD><xsl:value-of select="ait:Object/ait:ID"/></TD>
  <TD><xsl:value-of select="ait:Object/ait:Description"/></TD>
 </TR>
</xsl:for-each>

   </TABLE>
   </BODY>
   </HTML>

 </xsl:template>
</xsl:stylesheet>
```

To run the solution:

1. Save the mismatched topic to XML, once for English and once for Japanese.

2. Use an XSLT transformation tool to transform the XML files to HTML reports. There are free transformation sites. You select the XML, the XSLT (the code above that you saved as an XSLT file)and then click Run (or similar). The result is the HTML report.

3. Open the results in a spreadsheet. English will be one spreadsheet file and Japanese will be another.

4. Sort the data of each sheet by GUID.

5. In the sheet with the most rows, add a column with this function:

    ```
    =VLOOKUP(<cell with value to lookup in other file>,<table
    that might have that value>, <index number of column in
    that table that has the value you want>)
    ```

    Example: You want to see if the GUID in the JP output is in the EN output. If it is, get the EN GUID (where test.htm is the name of the EN sheet and the EN ID is in column B):

    ```
    =VLOOKUP(A2, [test.htm]EN01!$A:$C,2)
    ```

6. If the Descriptions are the same in the EN and in the JP, this topic is not causing the issue. If the Descriptions are not the same, your Translation Memory is out of synch:

    ```
    =IF(VLOOKUP(A2, [test.htm]EN01!$A:$C,3)=C2,"OK","out of
    synch")
    ```

5.7.2. Example: Fix PDF Bug

Here is an example of a Word macro that puts an empty string in the document author, subject, and other properties to carry over when the DOC file is saved as PDF. This fixes an old PDF security bug: if a property is not given, even as empty, anyone can add it with their own value.

To use this macro, enter your company name in the value of **myCompany**.

```
Sub FixDocumentProperties()
 myCompany = "your company name"
 With ActiveDocument
   .BuiltInDocumentProperties("Title")=         .BuiltInDocument
Properties("Title")+" "+ .BuiltInDocumentProperties("Subject")
+" "+   .BuiltInDocumentProperties("Keywords")
   .BuiltInDocumentProperties("Company") = myCompany
   .BuiltInDocumentProperties("Comments") = "Copyright" +
myCompany + ", All Rights Reserved."
   .BuiltInDocumentProperties("Author") = myCompany
   .BuiltInDocumentProperties("Subject") = myCompany +
"technical publication"
   .BuiltInDocumentProperties("Keywords") = myCompany
   .BuiltInDocumentProperties("Manager") = myCompany
   .BuiltInDocumentProperties("Category") = myCompany
   .BuiltInDocumentProperties("Hyperlink base") = "www." +
myCompany + ".com"
 End With
End Sub
```

5.7.3. Example: Highlight Long Sentences

If you have a Controlled Language checker tool, you will not need this macro. It is nice for quick checks if you work in Word as your source.

```
Sub ShowLongSentences()
 Dim iMyCount As Integer
 Dim iWords As Integer
 Dim MySent As Range
 If Not ActiveDocument.Saved Then
    ActiveDocument.Save
 End If
'Reset counter
 iMyCount = 0
'Set max allowed word count in sentence
 iWords = 25
```

```
For Each MySent In ActiveDocument.Sentences
   If MySent.Words.Count > iWords Then
       MySent.Font.Color = wdColorRed
       iMyCount = iMyCount + 1
   End If
 Next
 Selection.Find.Style = ActiveDocument.Styles("Heading 1")
 Selection.Find.Execute
 Selection.InsertAfter ("Report" & vbNewLine)
 Selection.InsertAfter (iMyCount &" sentences longer than " & _
iWords &" words." & vbNewLine)
End Sub
```

5.7.4. Example: Quick Pagination

This macro puts a page break on a paragraph. If you set a button in your Word to run the macro, it will help you paginate quickly, without accidentally undoing or duplicating the style. This is actually two macros. One puts in the page break, and one removes the page break.

```
Sub ParaNewPage ()
  With Selection.ParagraphFormat
    .PageBreakBefore = True
    .SpaceBefore = 0
  End With
End Sub

Sub ParaRemovePageBreak ()
  With Selection.ParagraphFormat
    .PageBreakBefore = False
    .KeepWithNext = False
    .SpaceBefore = 6
  End With
End Sub
```

5.7.5. Example: Make PDFs Open with Bookmarks

By default, PDFs do not open with the Bookmark tab open. For long PDFs, this is a shame. This script is written in AutoIt (au3) with the SciTE editor. It works if you have a Word doc open and then click a button on the taskbar to run this script.

```
#Region ;**** Directives created by AutIt3Wrapper_GUI ****
  #AutoIt3Wrapper_Icon=C:\scripts\pdficon.ico ;
** save ICO file to be desktop icon to call this script
** and change the pathname
```

```
  #AutoIt3Wrapper_RES_Comment=puts in pdf bookmarks
  #AutoIt3Wrapper_RES_Description=create pdf, ask draft, make
bookmarks
  #AutoIt3Wrapper_RES_Fileversion=2.0.0.3
  #AutoIt3Wrapper_RES_Fileversion_AutoIncrement=y
  #AutoIt3Wrapper_RES_LegalCopyright=rochelle.fisher@gmail.com
  #AutoIt3Wrapper_RES_SaveSource=y
  #AutoIt3Wrapper_RES_requestsExecutionLevel=asInvoker
#EndRegion

; from Word run Draft macro, save Word as PDF, save PDF with
bookmarks

#include <Word.au3>
#include <Constants.au3>

MsgBox (0,"Word2PDF: Wait","Click in doc to save as PDF with
bookmarks...wait until done")
WinWaitActive ("[CLASS:OpusApp]") ; the class OpusApp is from
AutoItv3 window info - this is Word
Global $myDraft = MsgBox(4,"Word2PDF: Draft or Final?", "Do you
want a Draft watermark?")
If $myDraft = 6 Then DraftMarkme()
    Send ("!fy1") ;Send ("!") means press Alt. Alt+fyi = in
Word means Save as Adobe PDF
; if the overwrite question appears, say yes
If WinExists("[CLASS:OpusApp]", "Do you want to overwrite it?")
Then Send("{Enter}")
WinWaitActive ("[CLASS:AcrobatSDIWindow]")
Send("^d") ; ctrl + d
Send("!n") ; alt + n goes to Navigation option. If the Initial
View is open
Send("b") ; enter the first letter of the Bookmarks option to
select it
Send("{ENTER}") ;press Enter to select default, save
preferences, close
Send("^s")
Send("!fc") ;close the pdf
Send("!f1") ;open the PDF and click File > Recent #1
Func DraftMarkme ()
    Send("!lpm") ;alt L opens developer ribbon, then the p and
m open the macro choice
    Send("DraftMark")
```

```
    Send("{ENTER}")
EndFunc
```

The AutoIt code calls a Word macro named DraftMark. This is the macro.

```
Sub DraftMark()
    ActiveDocument.Sections(1).Range.Select
    ActiveWindow.ActivePane.View.SeekView =
wdSeekCurrentPageHeader
    Selection.HeaderFooter.Shapes.AddTextEffect( _
    PowerPlusWaterMarkObject96411093, "DRAFT", "Calibri", 1,
False, False, 0, 0).Select
End Sub
```

5.7.6. XML Reports

Many authoring tools are based on XML. If you can get to the XML, you can generate reports and quickly find information that is otherwise not available.

With AuthorIt, we created publishing profiles with the AfterPublish options to run CMD commands and select XSLT files. The examples here use the AuthorIt schema and commands. You should be able to port the concepts to other tools and schemas.

If you have an XML file of a specific schema, and an XSLT to create an HTML report, there are different ways to run the transformation. If your authoring tool does not offer custom transformations, you can find free tools in SourceForge. If you plan to do a lot of work in XML and XSLT, consider getting a professional tool. I felt comfortable with XMLSpy.

Example: Open Report in Excel

We run this batch code as a *post-publish action* in AuthorIt. You can use it in a larger script with a different authoring tool.

```
REM %~d1 is the drive letter of the publishing folder, necessary
REM if the authoring tool publishes to a drive this is not C.
REM The first command goes to the drive. The second CD
REM command goes to the path of the publishing folder. %1 is
the arg
REM passed by the publishing profile to the script. This first
cmd
REM goes to the drive.

%~d1
CD %1
```

```
REM the XSLT reports each give a different suffix. Change this
REM DIR command for each report, to call the suffix of the
report
REM the DIR makes a file with one line: filename of the report

DIR /B *.<ext>.html > reports.txt

REM the next line puts the line of reports.txt into the variable
REM "_MyFile". The /P is necessary

SET /P _MyFile=<reports.txt

REM the next line opens Excel to the report file

"C:\Program Files (x86)\Microsoft Office\root
\Office16\EXCEL.EXE" "%_MyFile%"
```

Example: XSLT to find Referenced Topics

In AuthorIt, if a book failed it publish, it was often because a referenced file was deleted. This XSLT shows the objects that are used in a book.

```
<?xml version='1.0' encoding='ISO-8859-1'?>
<xsl:stylesheet version="1.0" xmlns:xsl="http://www.w3.org/
1999/XSL/Transform"
xmlns:xs="http://www.w3.org/2001/XMLSchema"
xmlns:ait="http://www.authorit.com/xml/authorit">
<xsl:output method="html" />
<xsl:template match="/ait:AuthorIT">
 <HTML>
  <HEAD>
   <TITLE>AuthorIT XML Report - Find all Referenced AIT Objects
   </TITLE>
  </HEAD>
  <BODY>
<!-- Display table that shows topic object ID and any
referenced object IDs. The @ means to get value of attribute -->
   <TABLE>
    <TR align="left">
       <TH>In topic</TH>
       <TH>fref tag to this topic:</TH>
    </TR>

<xsl:for-each select="ait:Objects/*">
    <TR>
```

```
        <TD><xsl:value-of select="ait:Object/ait:ID"/></TD>
        <TD>
    <xsl:for-each select="ait:Text/ait:p/ait:fref">
        <xsl:value-of select="@id" /><br />
    </xsl:for-each>
        </TD>
    </TR>
</xsl:for-each>

    </TABLE>
  </BODY>
  </HTML>
 </xsl:template>
</xsl:stylesheet>
```

Example: XSLT to Output Comments

We used this as the basis of our diff report. As we wrote, we kept comments in the topics for questions to ask and for changes with names, dates, and justifications. The Comment paragraph was a style that we made. Its element was p, and its ID was 1811.

```
<?xml version='1.0' encoding='ISO-8859-1'?>
<xsl:stylesheet version="1.0"
    xmlns:xsl="http://www.w3.org/1999/XSL/Transform"
    xmlns:xs="http://www.w3.org/2001/XMLSchema"
    xmlns:ait="http://www.authorit.com/xml/authorit">
    <xsl:output method="html" />
    <xsl:template match="/ait:AuthorIT">
<HTML>
    <HEAD>
        <TITLE> AuthorIT XML Report </TITLE>
    </HEAD>
    <BODY>
<!-- Display List Of Objects table -->
<xsl:for-each select="ait:Objects/*">
    <xsl:for-each select="ait:Text/ait:p">
    <xsl:if test="@id = '1811'">
        <p>In
                <xsl:value-of select="ancestor::/ait:Object/
ait:Description" />
                (<xsl:value-of select="ancestor::ait:Objects/
ait:Object/ait:ID" />), from
                <xsl:value-of select="ancestor::ait:Objects/
ait:Object/ait:ModifiedDate" />
```

```
        <xsl:value-of select="."/>
      </p>
   </xsl:if>
   </xsl:for-each>
   <xsl:for-each select="ait:Text/ait:table/ait:tr/ait:td/
ait:p">
   <xsl:if test="@id = '1811'">
      <p>In
      <xsl:value-of select="ancestor::ait:Objects/ait:Object/
ait:Description" />
                (<xsl:value-of select="ancestor::ait:Objects/
ait:Object/ait:ID" />) , from
      <xsl:value-of select="ancestor::ait:Objects/ait:Object/
ait:ModifiedDate" />
      <xsl:value-of select="."/>
      </p>
   </xsl:if>
   </xsl:for-each>
</xsl:for-each>
   </BODY>
</HTML>
   </xsl:template>
</xsl:stylesheet>
```

Example: XSLT to Find Help ID

In a company that delivered a Windows-only product, creating a CHM was a cost-effective method. To make the Help open to a relevant page on F1, we put a specific ID in the Help topic. This ID came from a map file that R&D created, which was compiled with the CHM to make the Help work.

AuthorIt did not provide a way to search for Help IDs. We created this XSLT to see all the IDs of a book.

```
<?xml version='1.0' encoding='ISO-8859-1'?>
<xsl:stylesheet version="1.0"
   xmlns:xsl="http://www.w3.org/1999/XSL/Transform"
   xmlns:xs="http://www.w3.org/2001/XMLSchema"
   xmlns:ait="http://www.authorit.com/xml/authorit">
<xsl:output method="html" />
  <xsl:template match="/ait:AuthorIT">
    <HTML>
        <HEAD>
            <TITLE>AuthorIT XML Report</TITLE>
        </HEAD>
```

```
        <BODY>
            <TABLE>
                <TR align="left">
                    <TH width="10%">ObjectID</TH>
                    <TH width="10%">HTML Filename</TH>
                    <TH>CHM Alias</TH>
                </TR>
    <xsl:for-each select="ait:Objects/*">
                        <TR>
                        <TD>
                                    xsl:value-of select="ait:Object/
ait:ID"/>
                    </TD>
                    <TD>
                    <xsl:value-of select="ait:WebFilename"/>
                    </TD>
                    <TD>
                    <xsl:value-of
select="ait:HelpContextString"/>
                    </TD>
                </TR>
    </xsl:for-each>
            </TABLE>
        </BODY>
    </HTML>
  </xsl:template>
</xsl:stylesheet>
```

5.8. Taking Control

What is a Controlled Language, and why should I care?

5.8.1. The Beauty of a Controlled Language

Adopting a Controlled Language is the answer to many questions. If you work as a team, it enforces consistency. If you translate, it makes translation cheaper. If you want a standard to follow, it gives you compliance rules and vocabulary.

5.8.2. CL for Newbies

A CL is a set of rules for grammar, sentence length, and vocabulary. The rules define a subset of the natural language as allowed. Specifically, many CLs limit the verbs you can use.

For example, it is correct in English to tell a user to enter input in a field with these verbs: input, enter, provide, type, do, perform, make, change, paste, write, tap, insert, add, amend, modify, dump, expound, configure, or tweak.

You can probably find more, but as I write this, I'm on the beach without Internet or a thesaurus. And I think you get the idea.

If your reader is a non-native English speaker, an inconsistent verb for this one meaning is confusing. At the least, it will slow down the user. And if this non-native speaker pushes your document through a machine translation, such as Google Translate, it will not be much better. The translation will not work well for all the words, and the user will wonder why you chose one word over another.

Your doc: `When you have configured the IP address in the A field, type your username in field B.`

User: *Configured?* What do I need to do more than enter the address? *Type* means to put in a class. Maybe that means user role? What roles are valid?

Now try this:

```
1. In A, enter the IP address.

2. In B, enter your username.
```

This is basic tech writing, isn't it? Most style guides even have a section of phrases to not use.

Example:

- describes how to perform X => explains

- in order to => to

- in case of => if

A CL makes this a standard that writers can apply to every task. It is more than a short list of guidelines. A CL has a well-defined vocabulary to be enforced. A style guide will have: *Write short sentences.* A CL will have: *Sentences in steps must be 20 words or less. Sentences in conceptual paragraphs must be 25 words or less. Phrases in parentheses are counted separately. You cannot have more than one parenthetical phrase in a sentence.*

Do not try to memorize this yet. This is just the start.

5.8.3. Benefits of a CL

We looked at two benefits. Let's make a clear list.

- Consistency over time and authors. Your team all write with the same words and grammar. They can re-use each other's sentences and topics. They can share the load of one book. And if you are the only writer, you can more quickly return to a dropped project.

- Productivity. Writers who truly internalize the CL rules tend to write faster. The increased re-use also makes for faster work.

- Clarity. The standardized vocabulary prevents ambiguity. Non-native speakers can read CL writing faster, with greater understanding, and with more correct usage. Often, translation is not necessary.

In MEGAComm 2018, I ran a short test with the participants of a session that showed readers of a CL-written procedure follow procedures with more accuracy than readers of uncontrolled writing. I published the results in a LinkedIn article: https://www.linkedin.com/pulse/results-rochelle-fisher/

- Cheaper translation. When legal or sales require localization, translation of CL writing is much cheaper. Shorter sentences with consistent words give more perfect matches (free) between a translated sentence and new sentences, and more 100% to 80% matches (much cheaper). Translation results are more consistent and generally better.

- Easier to read for native speakers. The idea of a CL is not minimalism. Its purpose is not shorter docs. Its purpose is simpler communication. Minimalist writing can be difficult to read.

In short, writing with a Controlled Language makes your documentation better, faster, and cheaper.

5.8.4. You Cannot Escape

If you think that translation has nothing to do with your work, think again. If you publish to HTML, most browsers offer a right-click > Translate option. If you publish to PDF, most users can paste your words in Google Translate. Machine translation is a fact of life.

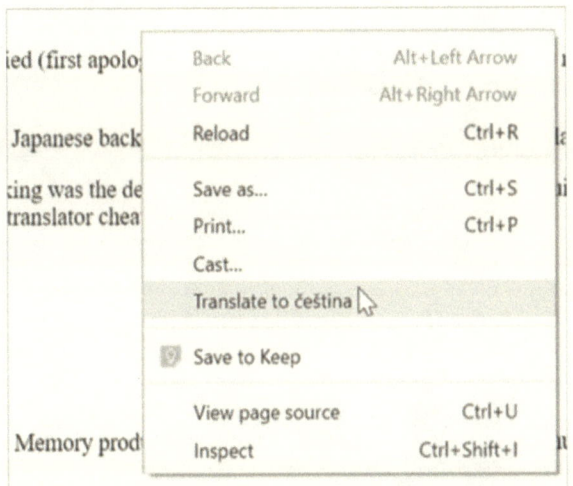

In February 2018, a Sales Engineer in Tokyo said he wanted the guides translated. I replied (first apologizing for the long answer) that such a project required a Localization Manager and project management, and I explained why. He answered that there is Google Translate.

In response, I did a snooty thing. I put one procedure in Google Translate and pasted the Japanese back to him, with "Is this really a reasonable translation?" I was shocked by his reply.

The decision was, "No. We need a human translator." No surprise there. What was shocking was the detail, "About 80% is good. The other 20% requires work." Just the year before, we used Google Translate to test translation bids. If the Japanese translated back to understandable English, we knew the translator cheated.

This really happens.

If you go with human translation, quality control is a must.

In just one year, the Google AI and crowd-sourced contributions to Google's Translation Memory produce 80% of a text without more work. That's huge!

Of course, the original English was in a CL, so let's get back to that.

5.8.5. Overview of Commercial CLs

You can choose from Controlled Language standards that are created and ready to use.

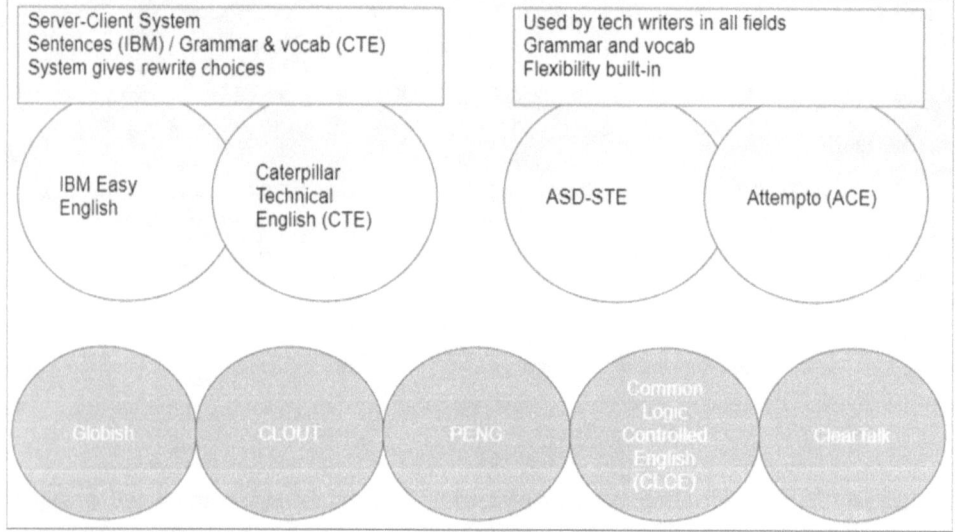

ASD-STE 100

This CL for Simplified Technical English is the one in vogue. The tekom organization offers training in STE in its technical writer certification program.

I should write tekom and not *the tekom organization*, but the correct capitalization is lower-case. I find it difficult to start a sentence with a lower-case character.

See tekom.org to learn more.

The STE documentation is easy to learn. It is easy to use. To see if a word is allowed, search in the PDF for the word with an open parenthesis: "*word* (". Words in all-caps are allowed. Words in lower-case are not.

The grammar rules are simple and helpful. I found the sentence count rules above what we needed. We were able to accurately check for compliance only for a few of the rules. The creators of ASD-STE are against automatic checker tools. They want writers to internalize the standard. But Acrolinx makes a good checker that integrates with AuthorIt.

The downside of STE is that it was created by Boeing for the aircraft industry. It has a lot of words, both allowed and not allowed, that you will never use in software documentation.

Caterpillar Technical English

CTE was a model for STE. Its vocabulary is for the specific purpose of caterpillar heavy machinery. STE has rules that to let you add Technical Nouns, but CTE is strict. Its terms can be checked by the CAT Automated Machine Translation machine.

A big difference for writers is that CTE is meant to be automatically checked, as you write and when you are done. This can be a big help when you first commit to writing in a new way. But it can be annoying when your authoring application asks you to choose a meaning for every word it marks as ambiguous.

The Cons of a Ready-Made Dictionary

After you learn to use a controlled language, the biggest disadvantage is that the ready-made CLs are built for a purpose. Even with the built-in flexibility of ASD-STE, you have a long list of words that you will never use.

5.8.6. Creating Your Own CL

You can avoid many of the issues of commercial controlled languages if you can create your own.

1. Decide on the verb tenses to allow. Avoid verb phrases, such as are needed for passive and perfect tenses. Imperative, present simple, and past simple. Try to keep it simple.

2. Decide on sentence length to allow. Different sources give a 25-word limit. Try to keep it shorter for steps. Decide if you allow breaks in sentences, such as parentheses or hyphens.

3. Decide on syntax for lists, table cell text, and annotations. For example, you can make a rule that notes for Warning annotations always start with the condition or always start with what to avoid.

4. Create the vocabulary (see the next section).

5. Develop a checker system. This can be a tool that your organization acquires or develops, an editor, or a peer review methodology for your team.

6. Communicate use of the CL as an innovation with change management. and refresh knowledge of your CL and enforcement during team meetings.

Creating Your Vocabulary

This is the most important part of a CL. It takes about an hour to work through 20 words. For a technical vocabulary of 5,000 words, expect to complete the vocabulary in about 32 man-days. If that sounds like too much time for your team, take an existing CL, such as STE, or get a consultant to parse your documentation for you.

If you want to do it yourself, here is one solution with a parsing script. The script lists unique words and gives the number of instances of each word. This example is a Perl script, but you can rewrite it in the code you prefer. It's a simple script. If you don't code or don't have time, you can ask a developer or student to do it for you.

To create the word parser script:

1. Download and install Perl or Python.

2. Select a document that represents your products.

Do not think on this too long. You can parse a number of documents, if necessary.

3. Copy this text to an editor:

```perl
use strict;
use warnings;

my %count;
my $file = shift or die "Usage: $0 FILE\n";
```

```perl
open my $fh, '<', $file or die "Could not open '$file' $!";
while (my $line = <$fh>) {
    chomp $line;
    foreach my $str (split /\s+/, $line) {
        $count{$str}++;
    }
}
foreach my $str (sort keys %count) {
    printf "%-31s %s\n," $str, $count{$str};
}
```

This script is on https://perlmaven.com/count-words-in-text-using-perl.

The author is Gabie Saab.

I tested it and confirm that it gives predictable results.

If you would rather use Python, here is a script that I made:

```python
import re
import string
import sys

frequency = {}
document_text = open("dict.txt," 'r')
text_string = document_text.read().lower()
match_pattern = re.findall(r'\b[a-z]{4,15}\b', text_string)

for word in match_pattern:
    count = frequency.get(word, 0)
    frequency[word] = count + 1
    frequency_list = frequency.keys()
for words in frequency_list:
    print (words, "|," frequency[words])
```

4. Save the text as a file with the extension: .pl (Perl) or .py (Python).

5. Start **cmd** and run:

```
{perl | python} fullpath/yourscript.py fullpath/yourdoc >
output.txt
```

89

Example:

```
python C:\myscripts\wordCounter.py C:\docs\currentdoc.docx
> count.txt
```

The angle bracket (>) is a redirect. It tells the processor to output to the file and not to the cmd window.

6. Open the output in a spreadsheet application.

   The result shows the words in the first column and the count of each word in the second column. Make sure the spreadsheet sees the count as numbers, and not as text (see that the count goes from 9 to 10, and not 1, 10, 2).

7. Add a column for **Part of Speech**.

   Each word is used as one part of speech in your writing. This is an important attribute to enforce disambiguation. For example, if you define *type* to mean *class*, but you do not set it to be used only as a noun, writers can use it as a verb and confuse readers between *classify* and *input*.

8. Add a column for definition. Each word must have one specific definition.

9. Optional: add a column for good example and a column for bad example. This is how STE does it. I find that helpful for matching definition and usage.

10. Schedule the work.

    Divide the number of results by 20. That is the number of hours of basic work. Add half that result to tweak the end results with a second document parse. Add one or two days to get started and one or two days to close this project. Set a deadline. Now you know when to start, how much time to spend on this every day or week, and how many writers to assign.

    For example, with 1000 results, you need 50 hours + 25 hours + 48 hours, or 16 full days. If your deadline is the end of the year, you can complete it with two hours a week for two writers. Or you can assign one writer to complete this in less than a working month.

11. Set the template for the dictionary.

    Decide how to mark words as non-allowed. Maybe the simplest way is to delete them, but that will make it more difficult to check the future documents. In STE, non-allowed words are lower case. Instead of the definition, it shows allowed alternatives. Or you can move non-allowed words to a different tab or worksheet on the same spreadsheet.

12. Go through the words. For each word:

    a. Look at the original documents to find the most common part of speech and definition used by your team.

b. Look for synonyms in the list. Do you really need "alerts," "messages," and "notifications"? If each is a unique definition, you can keep them all. If they are used to mean the same thing, choose one to be allowed. Do not allow that others.

c. Optional: compare your word with the dictionary of a ready-made CL. What does STE or Attempto say about this word?

13. Review the allowed words. Can you cut it down more?

5.8.7. Challenges of CL

With all these wonderful benefits, why doesn't every documentation team adopt a CL? For the same reasons your heart is squeezing right now. Do you really want to get into a conceptual innovation that you know nothing about?

Cost

If you choose a commercial CL, it will cost something. I don't know how much it costs to get CTE and its client-server application, but I can guess it is in the thousands or tens of thousands of dollars.

It does not matter which you choose. To follow through from idea to implementation, you must get professional training. One to three days is enough to start. There is a money cost and time cost that your organization can probably afford.

There are more time costs. You must continue to refresh your team in its use. To keep your vocabulary up to date with agile products and industry changes, you must put in time. In one team I managed, we went over the rules every team meeting for at least ten minutes, over the space of two years. We talked about words we wanted to add or replace, and we spent about an hour a week on research.

"I want to use the word *however*."

"If you can find an instance where it is necessary, let's discuss."

Next week...

"I searched all instances of *however* in our documents. I cannot find one instance where it is necessary. I agree that it not be in our vocabulary."

If you choose to create your own CL, it does take time.

Enforcement Issues

It is difficult to enforce compliance. You may have a writer who did not buy in to the idea. Other writers have habits that are hard to break. If you have a checker tool for complete documents, most writers will not have the time or inclination to change the documents for hundreds of comments.

Checker tools are never 100% accurate. I never used a commercial checker tool, but others have,[6] and they say the same. Usually I ask developers in the company to make a Perl or Python script. At one time, I manually checked the writing of more than five authors over a month, against the checker tool reports. There was an average of 60% false positives.

```
Found in: The message shows.
Term: Message
Rule: Not allowed as a verb, only as a noun
```

(In this example, *message* is used as a noun. But the word was hit by the checker tool.)

And there were hundreds of false negatives. One that I found so often I stopped counting was *have*. It is allowed as a verb, but present perfect is not allowed. The checker tool did not find grammar rule violations.

Native Speaker Dislike

In one company I worked for, we sent a survey for document usability to users. The first survey was in 2008, for documents written before the team adopted STE. The second survey was in 2012, when most of the documents were in STE. The grade given by non-native speakers was much higher for the new docs: 4.5 out of 5 (up from 2). But the grade given by native speakers was lower: 3 out of 5 (down from 4).

Native speakers felt the new style was dumbed-down. They felt a loss of aesthetics and nuance.

---

[6]Abel, Scott. Choosing the English That's Right for You.

Overcoming the Challenges

- **Cost** – Show the value.

  - You will see a return on investment after the first subsequent translation. The first translation will be expensive (average translation to one language for a 200-page document is about $20,000). The translation of the next version will only be the difference in text. Similar sentences (for example, the same syntax with one different noun between them) are cheaper than new sentences. Repeated sentences are free (perfect matches). After three translations to Japanese of an STE-based document, the cost of translation went from 20K to $200.

  - Run a user survey. You will see that users give higher grades for clarity and ease of use. Personally, I don't care that sales and partners find it too dry. I'm a customer advocate, not a marketing writer.

  - Encourage users to rate online articles. See if you get higher ratings for CL-based articles.
  When the value is clear, schedule time to update the vocabulary, review rules, and edit documentation for CL compliance. The time required is part of your team methodology.

- **Enforcement** – Keep the buy-in and commit to a checker methodology.

  - Make sure your team is committed to the CL. Share the value information that you collected. Be patient.

  - Choose a method or tool for checking compliance. You can hire an editor, buy a tool, have a developer create a tool, start peer reviews specifically for the language. Or maybe you can come up with a different idea.

  - Keep violation numbers of authors' checker tool results or editor comments. Track improvement and issues. Help your writers find solutions to issues. Celebrate improvements.

- **Native Speaker Dislike** – First, find out if the negative feedback is from real users or from partners and resellers. The best feedback I got came as negative for the new style. It was from a tech support engineer who made it a practice to read new versions of the product documentation cover to cover. He said the new book was 100 pages shorter than the previous. It must be missing information. We looked at it together. No, no task was removed. We did find that we were missing end-to-end tasks and complete use cases. But that was something we were always missing. Do not explain how to read your docs. But use internal feedback as a question to investigate. When real users complain, that is a call to action.

In the end, technical writing is not creative or amusing. Often, it is not interesting. It is not meant to be. The purpose of technical writing is rapid information retrieval. If readers think the writing is "dumbed down," they are reading it. We make the docs "smarter" by adding more relevant information, not by wrapping the same information in more complex writing.

5.9. Managing Localization

Translation is a project that requires proper management of resources, time, budget, and quality criteria. Without a plan or tools, you will pay too much for sub-quality output.

Let's get you started with the terms that you will see when you first get a project requirement for translation.

L10N = Localization (L - ten letters - N) is the project area that produces deliverables for a special locale. This often, but not always, includes translation. If you have a large customer in the UK, you can tweak the spelling and grammar from the generic American English to British English. If you want to go full-out, you translate, change colors, replace photographs and screenshots, re-do diagrams, and so on. For example, Uber in India changed their first photograph of a woman getting into a car with a male driver, to a female driver and passenger. They thought they had localized with photographs of local people, but market research showed cultural issues with their first try.

If your product GUI is translated, you must do the minimum of L10N for your translated documents - replace English screenshots with screenshots in the target language. If the GUI is not translated, but the document is, make sure there is a screenshot for each step, with a pointer or shape around the control mentioned in the step.

T9N = Translation (T - 9 letters - N) is the conversion of text from a source language to a target language. From your point of view, this is creating the translation kit and managing the translation as a project.

I18N = Internationalization (I - 18 letters - N) is the preparation of the source code of the product and the writing of the document for L10N and T9N. The developers must be able to parse GUI strings from the code. Then, you update your team procedures for I18N:

- Apply a strict style guide or controlled language
- Apply team methods for effective re-use of paragraphs and sentences that are often repeated.
- Apply a graphic rule that replaces callout text with numbers or symbols and a style guide for a legend or table that explains the numbers or symbols.
- Apply team methods to set translation states on components: In Progress, Approved Src, Sent to Trans, Translated, Changed Src.

G11N = Globalization (G - 11 letters - N) is Translation, Localization, and Internationalization. We should actually use G11N as the generic term, but translation providers use L10N, to promote their services, and LM (Localization Manager) is an industry term. We will use L10N for everything in this chapter.

Localization Manager is a half-time to full-time position, depending on the requirements of the organization. The first tasks of the LM are:

1. Define quality benchmarks.

2. Create a DNT and a Translation Glossary.

3. Create procedures for consistent and cost-effective translations.

4. Acquire an LM tool and learn how to use it.

5. Create a project plan to determine which version to translate first and how long it will take to deliver a translated result with the required quality.

5.9.1. Human or Machine?

Usually, a professional translator is necessary. Look for a local, native-speaking technical translator. This must be done correctly.

1. Get bids from at least three companies.

2. Give a pilot chapter to translate.

3. Review each for quality.

4. Decide on the company that best meets your quality benchmark at a price you are willing to pay.

5.9.2. Keeping the TM

To create a TM from existing docs (alignment), use a tool and then have a native translator make sure each line is correctly aligned.

Make sure you keep the TM as a backup. You can ask your vendor to send it to you every year. It is a backup for the translators. If you have a TM, you do not have to set the state of each topic and lock translated parts to not send because a good vendor will not charge for 100% match.

But you should have states on topics, to make sure that you send only the delta for subsequent translations.

5.9.3. Linguistic QA

Linguistic QA is not just translation. It is also context in the GUI. All translators and QA must see the GUI.

1. The strings are translated.

2. The translator reviewer sees the strings in the GUI and makes corrections for context. If you cannot create a demo GUI quickly, a native speaker should be with the

developers and change the strings on-the-fly to fit the context and space. This can take a full day, sometimes two.

3.   Send the translated strings to Sales and Solution Engineers. They know what their users and potential users expect. Remember to include this time in the schedule!

4.   The LM does a sanity check: is every string translated and does it fit in the provided space? When you pay for translation services, you pay for two: one to translate and one to review. You still want an employee, usually an SE, to review the translation.

Linguistic QA tests for these issues:

- Formatting errors - spacing is different for each language
- Grammar
- Punctuation
- Spelling
- Untranslated words
- Non-compliant terms

5.9.4. L10N Project Scheduling

First, make sure the localization project should be done. Do not waste time and money on something that will not sell because the competition did it better already. Before you begin, look at what the competition offers. If one of your main competitors already offers a translated GUI with similar features to your product, it might be the correct decision to let the competitor have this corner of the market.

The LM requires planning time to answer these questions:

- Are there design changes for globalization?
- Do you have, or can you get, analysis of target local markets?
- What is the priority, according to Sales, of the languages for translation?
- Are there government regulations in the locale to consider?
- Is the product ready for localization?
  During software development, the LM and the project owner must make sure the product is created to handle multiple languages. It must be language and region independent. The UX expert and the front-end developers must consider the code itself (ability to parse out strings and ability to toggle between sets of strings by locale) and the design of density, fonts, images without text, colors, spacing, button expansion. The LM must be able to tell the translators how much space there is for text on each II control. Dates, currencies, units, addresses, phone numbers are different between locales. The GUI must be

prepared to handle this. The localization-ready GUI must be tested in the source language and in a target or pseudo-language. These changes in the code can create unexpected feature issues.

- Will localization, to the first target language or at all, be cost-effective?

To determine this, the LM requires an overview of the project, estimated number of words, required deliverables, and deadlines. The cost of translation, on average, is $0.20 a word. An average admin guide is about 20,000 words. Thus, the cost of translation for the guides - only the guides, not the GUI - will have a base of $3,400. If your team writes for translation, there is re-use of text. You, or an editor, must review the guides before the LM sends the translation job, to make sure re-use is done consistently and with more oversight. If you do not have a Localization Management tool, the translators will charge for Translation Memory (TM) management and DTP, which are things you can control. If you do not get the tool, and you change translation vendors, the cost for the first **and for subsequent** translations can be up to $20,000 USD. If the LM handles the projects correctly, the first translation of a complete guide will be $10,000 - $20,000, and it will be cheaper for each update translation. There are Translation Management System (TMS) tools you might consider. These have a higher cost of initial investment and time for maintenance. Do not get one unless you have a volume of translation projects to justify it.

When the LM has the answers to these questions, it is time to create a schedule. Always add 15-20% of project time for translation. Translators do about 1500 words a day, when they actually start. It takes an average of one working month to translate a complete guide. You must also add days before and after (on average, 5 before and 5 after, to be reduced on subsequent translations), for the creation of the job and the kit, initiation communications, post-processing, debugging, and delivery. Include time for Linguistic QA. There should be test plans, and the LM should set up communication channels between the QA engineers and the translators, to fix issues that the translator can fix, or between the translators and the developers, to change the product when necessary.

5.9.5. Translation Kit

When you send a job to translate, include the full **translation kit**:

- Content to translate, in a format translators can use.

- Screenshots of the GUI in the source language and if possible, of a translated GUI.

- Instructions on how to access the product.

- TM of the strings. This helps with one-to-one translation. One string is translated in the same way on the GUI, in the docs, in the training, and in all your deliverables.

- TM of the documentation.

- Dictionary of the most important, the most common, or industry-standard terms, with their definitions, in the source language.

- Translation Style Guide: expected tone, formatting requirements, where to find more information, best practices in handling the project (if you have any), mock-ups and demo versions.

- Translation Glossary. This is a list of words, such as the current values of variables and basic terms used often in the documents, with the translation of the word in each localized language. If the GUI will be translated, add all the strings to the Glossary. That should take two to five days, but it can take longer if the GUI strings are not consistent and if the code is not set up for translation (cannot be parsed to see all the strings in one file).

- Do Not Translate list (DNT). Start with the list of variables because it includes company and product names. Send the list of variables - resolved to their latest default text - in a text file to an SE in the target country. Ask the SE which words he or she wants to see translated and which to keep in English.

 If you do not have variables for all the product names, you will spend hundreds of man hours when a VP decides to rebrand. And if you have variables for the company name and acronym, you can deliver an OEM very quickly.

Use the same drive or wiki for your localization plans, kits, bids, purchases, and results.

If you translate to more than one language, make the Glossary and DNT in a spreadsheet. The full list of source language words may have different decisions from the different SEs. For example, the Tokyo SE says that *firewall* should stay in English, but the Buenos Aires SE wants it translated to *barreira*.

6. CREATING A STYLE GUIDE

Consistency in formatting and style implies accuracy in content. Inconsistency implies inaccuracies. Worse, the linguistic signs of format and language can confuse readers if used incorrectly or inconsistently. Even if you are the only writer in an organization, you should have a style guide. If you are in a team, or leading a team, a style guide is a basic necessity.

6.1. The Joys of Consistency

The primary concepts to know about a style guide:

- It must be created.
- Writers must buy-in.

So, you have the Chicago Manual of Style. Or Microsoft Manual of Style. Or some other 500-page style guide. Cool. I dare you to use it. Go ahead, try! Bah! It has nothing to do with your corporation and its needs. No one reads it cover to cover. (Yeah, you small number of tech writer geeks like me, who read these books for fun, we don't count in this.)

You must create your style guide. Its headings must make sense to your team. It must be relevant. It must be usable. It should be dynamic. It must be searchable and easily accessed. Hm. I make it sound like it should be on a wiki. That is up to you. It works for me, but use whatever tool you like.

Your style guide explains how writers capitalize and punctuate different types of information, the word to introduce specific types of information, how to keep metadata to work better as a team, and so much more. Tools change, but your style guide is about basic writing decisions and reminders. Maintain your style guide separate from procedures that depend on specific tools.

6.2. Suggested Headings

Here are some headings to consider for your style guide. You and your team create it together. My notes and recommendations are in italics. Examples are in monospace.

- Foundation Image (vision, values)
  From the core values of the organization, the vision of the CEO, or your vision for the team, everything in the style guide must uphold the foundation image.
  ```
  Org vision: Provide easiest to use product >> Assume
  audience is lowest-tech level.
  Org vision: Global leadership >> G10N Rules.
  ```

- Deliverable Definitions (output types)
  See also: *More About Deliverable Definitions*
  For each output you deliver, define the purpose, basic outline, intended audience, and unique requirements.

- General Style Rules
  ```
  Do not use conitemizedlistactions
  Use Corporate colors, logos, and fonts
  Use the Oxford comma
  ```

My special request for your style guide:

Please use the Oxford comma! It makes docs easier to understand and less ambiguous. For example, "Download, install, and configure," means "Download it. Install it. Open it. Set the configuration properties." But "Download, install and configure," implies the only or primary configuration is during the installation wizard.

- Metadata (Why, when, and how to add internal comments)
  ```
  Use the remark tag to enter comments. Enter your username,
  date, and name of the SME.
  ```

- Filename conventions (Syntax for filenames and paths)
  ```
  CoShortName_Version_GuideType
  ProductNameVrs_WindowN_Screenshot
  ```

- Bugs & Fixes (How to collect data and write approved RNs)
  ```
  To collect known limitations, edit this JIRA query for the
  current release...
  Do not use these phrases: "Low performance", "Dead on
  arrival"
  ```

Open bugs are often in the Release Notes. Consider separating the bugs from the other parts of the RN. If the RN has information that does not change after the day of release, and if you must continuously push a new revision for updates - new bugs, changed text to published bugs, fixes - from R&D, you open your deliverables and your users to mistakes. Best practice: keep the bug list online in HTML format and do not publish them In a PDF or other file that can be obsolete by numerous revisions in a few weeks.

- Re-use Model (How to chunk information to re-use parts efficiently)

  ```
  Smallest manual chunk is one topic.
  Do not chunk phrases or words that are used often. Use
  variables.
  ```
  See also: *Best Practices for Re-Use*

- Procedures (This is the money shot)

  See: *More about Procedure Styles*

- Character Styles (Information objects that get specific styles)

  Do not go into detail on what the style looks like in the deliverable (for example, "bold"). That is for the templates, which can be changed easily for different media.
  ```
  Text to input and Filenames are in style InputText
  Text on the GUI is style GuiLabel
  Use "Click <cs>X</cs>" and not "Click the X button"
  ```

- Tables

  ```
  For lists of ITEM - DESCRIPTION do not use bullets, use a
  table
  Always have a Heading row and set it to style TableHeading
  ```

- Offset text (Styles and sitemizedlistucture for note, important, tip, warning)

  Make sure you have style names for the type of information you use. If you output to HTML and use styles or classes for offset text, you can conitemizedlistol the look and feel of each with a CSS that you write once and update easily.
  ```
  Text prefix Note and Caution, icon for tips and best
  practices
  Use style Note, Caution, or NoteOther
  Write the condition first, then the result or action,
  separated by a comma
  ```

- User interfaces (GUI, CLI, API, AR)

  Use industry-standard conventions and make sure everyone uses them consistently. You can write these in in a chunk to re-use in every deliverable. Make sure a user can easily see the difference between a command to enter and a variable to change.
  ```
  Write a result: "the X window opens" and not "in the window
  that opens" in a new step
  ```

Prepositions: "from the menu, in the window, in the field, on the widget"
CLI variables are italic

- IP Networks
  See: *More about IP Networks*

- Links and Cross References
  All cross-references are live (in PDF and HTML)
  Do not create a cross-reference to a topic in a different deliverable
  Use "see: <link>" without a period for website links

- Lists
  Everybody has a different idea on how to capitalize and punctuate lists. It is all a matter of style. Pick one. Be consistent. This is for all deliverables in your company. Team agreement and buy-in are paramount.
  Items do not have end punctuation, unless one item is more than one sentence
  If one item has a period, all items in that list have a period

- Examples
  At least one procedure in a section must have an example
  All new features from version x must have an example
  See also: *More about Examples*

- Graphics
  Rule to never use shots depends on the delivered media and your resources compared to your commitments. If you do not have the resources to keep screenshots updated, you can say: never. Be prepared to fight for that (and be prepared to lose often).
  Use number callouts with a text key and explanation
  I recommend you keep text out of graphics, which are hard to update and costly to translate.
  Use the file name convention and store in directory X:\Graphics
  Source and final graphic have the same name
  Is there anything worse than creating a complex topology graphic again because the source file cannot be found?

- Glossary (When to include, title, placement, definition syntax)
  See: *Creating Glossaries*

- Print

For those who still deliver mutilated tree carcasses
```
Total number of pages must be divisible by 4
Always start a chapter on odd page number
Deliver to printer with crop-marks
Options and cost of printing (tones or colors, paper
weight, cover material, copy bundles)
```

- Security and compliance requirements
```
Set PDFs to now allow copy and paste
Deliver KB articles for Support-required workarounds to
Internal only sections
```

- Storage and Maintenance
```
CCMS auto-backup every 3 hours
Save backups to corp Drive
```
Do not rely on magnetic tapes only. They are easily corrupted and desitemizedlistoyed.

6.3. More About Deliverable Definitions

Standard deliverable types are: Administrator Guides, User Guides, Release Notes, Getting Started Guides, Quick Start Guides, Knowledge Base articles, Troubleshooting Guides, In-Product Help, Chat-bot scripts, How To videos, Linux man pages, and soon Augmented Reality overlays.

The definition of each is different for different organizations. Your definitions must fit the purpose and audience of each deliverable type you offer.

Define what you do and what you do not do as quickly as possible. Translations, white papers, data sheets, tutorials, FAQs, marketing, and F1 Help each require a different skill set. Each is a different deliverable with expected content, style, and formatting. If you do not know how to create a non-required deliverable, do not commit to creating it. Then learn about it as quickly and deeply as possible.

6.3.1. Outlines

For each deliverable, add a basic outline to your style guide. This outline has front matter (copyright), back matter (contact information), first chapter (choose "Introduction", "Overview", "Preface", or similar, and make sure it is consistent for all deliverables of that type), default chapters or types of chapters, and whether you use appendixes and an index.

In my opinion, the Index is dead. There are so many automatically made indexes, which add nothing to a PDF or HTML full-text search, that users do not look for information there. A well-made index, that gives synonyms not found in a search, takes (on average) an hour for every ten pages or topics of the deliverable. If the users do not trust indexes and thus never use them, you will never see a ROI on that time. And if you translate to non-Latin languages, the sort of the Index is a fail point.

Decide if you have a minimum number of pages for a Table of Contents to be added.

6.3.2. Release Notes

Pay attention to the most important deliverable for most companies: Release Notes. Make sure the outline of the Release Notes is well-defined.

- Information in this document must not be duplicated in other documents. If you single-source a topic of requirements to multiple documents, you do not always re-publish released documents when the requirements change. A user can get an out-of-date requirement list and not know it is incorrect. Or you can easily deliver multiple documents for one release with conflicting basic information. This isn't the end of the world if the unsynchronized data is in a word or two in a procedure, but it could mean the end of a job for a decision-maker who made a purchase based on the requirements you published.

- Pay attention to how your team writes the section for new features. That section gives the first impression of your company's offerings. If your first section of the first read document - the new features - is stable, easy to use, consistent, accurate, and error-free, that is the impression your users will have of the product itself.

6.3.3. Context Sensitive Help

Before you decide to have context-sensitive Help (F1 Help), make sure your users expect it. Discuss with your UX design team or Usability testers. If your users go to Google faster than your in-product documentation, minimize the impact of this documentation feature on your resources.

If you deliver context-sensitive Help, decide if it will define every control on a window, or if it will be best practices and troubleshooting.

Please do not choose to document EVERY control. Users do not tolerate their time being wasted with lines about the meaning of OK and CANCEL.

Can you create an outline for every topic in the Help without adding irrelevant sections? Make sure the procedure for creating Help is well documented for everyone on your team. This can be technical and complicated, and if done incorrectly, it can break a product build.

6.3.4. Single-Sourcing Guidelines

You can add a line about single-source between types of deliverables. But separate the deliverable definitions from the re-use model and styles.

6.3.5. Re-Use Best Practices

Re-use means to select a chunk of writing that is used as-is in different deliverables. After the first time the information is written, re-use is by reference, not text.

There are different models for chunking, re-use, and single-sourcing. Pick one, test it, and bring it in your style guide.

If you re-use chunks correctly, you get automatic updates (faster, more accurate) and cheaper translation.

Let's look at best practice for chunking size. Of course, the larger objects can contain nothing but smaller objects (a file can contain one sentence). AuthorIT vendors will say you can chunk sentences, if you use a topic of one sentence. And you can chunk all sizes with Word, if you save the smaller chunks in a file. In this case, we talk about the pros and cons the size of information objects according to their usual and most efficient use.

Note: I mention tools here. This is from my own experience and of course is not a complete list. Vendors might disagree with my understanding, so take the tool part as a guideline to discuss with tool vendors.

Chunk Size	Advantages	Challenges
Sentence Tools: Paligo	Quick reflection of GUI changes. Example: "Click OK" is written once and re-used 100 times. "OK" is changed to "Yes" in the GUI. You update the docs in 1 minute.	Writers stop using sentence chunks when there are too many possible matches. It takes longer to find the sentence than to write it. Some tools do not resolve variables. You cannot search for the sentence or see it to select.
Paragraph Tools: FrameMaker	Consistent descriptive text that promotes the organization's brand.	Changes do not always apply to every instance. If you are unlucky enough to miss the non-relevant usage, you create bad documentation.
Topic Tools: AuthorIT, Paligo, FrameMaker	Easy to find and to reference. Example: Drag a topic component to the book container.	Do not re-use complete multi-paragraph topics in one deliverable (unnecessary duplication). If topics are for smaller chunks, organize carefully. You have the disadvantages of the smaller chunks without tool-based features to help. If re-used in multiple deliverables of the same release, benefits come only in the next version.

Chunk Size	Advantages	Challenges
		If re-used in multiple deliverables of multiple releases, you will forget to re-publish released deliverables for the changes.
File: MS Word, OpenOffice Writer	Free. This is basically Topic re-use without a CMS.	Organization of files is usually not part of the tool. You will have to find a 3rd Party tool or methodologies to enable search and reference. Same challenges as in Topics.

6.4. More About Procedure Styles

Some companies see this part of the Style Guide as the most important. They will put in a full example. Writers can find the example easily and let it guide them when they don't have time to read and analyze rules for practical application.

If your user guides are filled with back-end specifications, you are writing internal documents that an engineer could write better. If you write section after section of the benefits of the product, you are a marketing writer. If your documents answer how to use the product, in a format that allows for rapid information retrieval, you are a technical writer. The easiest technical document to read, and to put to use, is all about the procedures. Step > result > next step.

I see different types of procedures:

- Workflows - Light on the details, lots of cross-references to specific steps

- Specific Tasks - One, two, three...ten, and you are done

- Alternatives - Tell the user when to choose one thing over another (avoid this if there is no difference in scenario or outcome)

- One Step - There's only one thing to do, but it must be done now

Do these types fit your needs? Do you have others? How to introduce the steps? For example, will every set of steps have a complete sentence, or a *To Do* statement to introduce it? What is the punctuation and basic style of the introduction? Will the steps be numbered sentences, with results in unnumbered next lines? Or will every step have a result on the same line as the instruction? If each step is one instruction with no result on the same line, will the step have punctuation? Will every procedure end in a result, no matter how obvious? Or will every procedure end on "Click OK."? Or will the end of a procedure be a list of What's Next procedures? If you are using component-based (topic-based) writing, do you want all procedure topics to have a heading and introduction?

You could have a style guide that says: *Every procedure has a gerund heading, then a style that outputs OBJECTIVE on the side, and the next style outputs PREREQUISITES, then INSTRUCTIONS and then step 1.* (Actually, your style guide should say that procedures have one heading, objective, prerequisites, and instructions section. Your tool creates the styles automatically.)

You could have a style guide that says nothing more than: *Procedures have numbered steps.* You can leave the rest of it flexible.

6.5. Much More About Tables

Many XML-based authoring tools handle tables differently and less-well than text. Sometimes the placement of the first column is uncontrollable. Sometimes the width is not aesthetic.

Do you set table widths? Your style guide can say that commands must not be broken, so the column with the command words is always as wide as the longest command. Or your style guide can say that text in a cell must be two lines or less, so the column with the description of a command is wider.

If your tables have visible borders, it is best if the last columns all line up. In English (left to right), it might not be important that tables all start at the same point, but it is important that they all line up on the right.

6.5.1. Power Up with a Table Macro

You must change table widths for different page sizes if you deliver PDFs or printables. You can run these macros on a Word (or compatible) document, to resize all tables.

To set all tables to the width of an A4 page (17cm, 480 points):

```
Sub TableWidth17()
    Dim objTable As Table
    For Each objTable In ActiveDocument.Tables
        With objTable
            .PreferredWidthType = wdPreferredWidthPoints
            .PreferredWidth = 480
        End With
    Next objTable
End Sub
```

To set all tables to a width that the writer sets (for non-standard page sizes):

```
Sub TableWidthOther()
    Dim objTable As Table
    Dim myWidth As Integer
    myWidth = Int(InputBox(Prompt:= _
        "How many points wide should the tables be?",_
        Title:="Non Standard Table Width",_
        Default:="1cm = 28.35 pnts"))
    For Each objTable In ActiveDocument.Tables
        With objTable
            .PreferredWidthType = wdPreferredWidthPoints
            .PreferredWidth = myWidth
```

```
        End With
    Next objTable
End Sub
```

6.6. More about IP Networks

When you write technical documentation for a product that builds, works with, or depends on networks or the Internet, you will have IP addresses. It is very important that you do not expose actual IP addresses to the general public. Use the IP address ranges that are reserved for documentation.

The Internet Engineering Task Force (IETF) drafts standards in Request for Comments (RFC) documents. When the standard is agreed upon, the document is done. The meaning of the name is lost. The important thing to know is that an RFC with a number and title is a standard you can use.

These RFCs explain the IP address ranges that are reserved for documentation:

RFC 3330 - Special-Use IPv4 Addresses

RFC 5737 - IPv4 Address Blocks Reserved for Documentation

RFC 3849 - IPv6 Address Prefix Reserved for Documentation

6.6.1. Summary of IPv4 Standards for Writers

IPv4 ranges are shortened with a slash and a bit number.

- If the bit number is 8, the first bit of the address stays the same, and the last 3 can go between 0 and 255. For example, with 100.0.0.0/8, the range is 100.0.0.0 – 100.255.255.255.

- If the bit number is 16 - the first 2 bits stay the same, and the last 2 go between 0 and 255.

- If the bit number is 24 - the first 3 stay the same, and the last 1 goes between 0 and 255.

For IP addresses of a vendor network (for example, to show the IP address of **example.com**), use **192.0.2.0/24** - 192.0.2.0 to 192.0.2.255.

To show two or three separate networks, these are the other ranges: **198.51.100.0/24** and **203.0.113.0/24**

Try to keep 1 as the last bit for default gateways.

Some network product vendors will change the purpose of the reserved ranges. If your company uses one or two of these blocks for the product or internal testing, use the other range for documentation.

6.6.2. Private Network Standards for Writers

If you document private networks (behind a router or gateway), these blocks are the only relevant IP addresses to use:

10.0.0.0 - 10.255.255.255 (10/8)

172.16.0.0 - 172.31.255.255 (172.16/12)

192.168.0.0 - 192.168.255.255 (192.168/16)

6.6.3. IPv6 Reserved Addresses for Writers

http://www.ietf.org/rfc/rfc3849.txt

An IPv6 address looks like a MAC address. IPv6 addresses are usually shortened to replace consecutive zeros with double-colons. The reserved range for documentation is: 2001:db8::/32. This gives the range as 2001:db8:0000:0000:0000:0000:0000 (written 2001:db8::) to: 2001:db8:ffff:ffff:ffff:ffff:ffff:ffff

Recommended style: replace consecutive zeros with double-colons and use lower-case letters.

6.7. More About Examples

Examples of use cases are awesome tools. Too often, we leave them for the end, and then do not have time to put them in. If you want examples, make sure this is part of the documentation project plan.

The more complex the procedure, the more likely users will look only at the examples. And the more often Support will get calls because the user followed the example as given, without changing it for local values. Make sure that a pressured user is very clear on what in an example is meant to be changed.

A good use case example tells a story. This is a possible outline for scenario examples: Introduce a user with a role. Explain the problem the user faces. Explain the solution through the product.

There are style guides that use examples as a means of promoting new features. Every new feature gets one or two complete scenario examples. After one or two versions, those examples are shortened or removed. In this way, the guides always give more space to the new features. It maintains an importance-by-space perspective of older features. Without this, features get documentation real estate by arbitrary factors of time available and whim of the current writer.

Be careful of usernames. Do not use names of celebrities. This can make your company vulnerable to lawsuits. And it can open your documentation to emotionally based connotations, which can hinder rapid understanding. For example, if you use "Dan Brown" in your use case, and a reader loves the well-known writer of that name, the emotion kicks in and comprehension slows down. For a similar reason, avoid using the world's most common personal name: "Mohammad" (with its variants "Muhammad" and "Mohammed"). Religiously charged names bring emotional connotations. For some readers, it is a form of blasphemy. Pick simple names that are easy to pronounce, do not have ambivalent meanings, and do not take up too much of the line when used in CLI commands.

6.8. Creating Glossaries

This section is about creating guidelines for a glossary of terms for guides. If the guide is less than one hundred pages, does not describe new technology, and does not include inline definitions, you do not need a glossary.

But if some of your deliverables are longer, the technology requires explanation, or your writers are defining each new term as it comes up, you need this.

6.8.1. Form and Function

All deliverables of an organization should have a consistent form and placement of the glossary. This is something you can let a team consensus determine.

- Should the Glossary go at the beginning or end? If you want the Glossary to plug new features in a marketing way, put it at the front. If this is for a new release of a stable product already in market for more than two years, you might prefer that the Glossary be at the end. If you put it in the end, make sure it shows in the Table of Contents.

- Should the title be "Glossary" or "Terms" or something else?

- Should it be a table or a list or an automatic generation from the tool (usually a 2-col page)?

  - If a table, should the columns have set widths, or should the term column be as long as the longest word for each deliverable?

  - If a list, are the definitions on the same line as the term or the next line?

  - If same line, what is the delimiter between the term and the definition?

  - If a two-column layout, set a maximum length for a term. Don't let the term run to a second line. That is icky.

The form you pick must enable alphabetical sorting. If you do not sort the terms, they are harder to read, and you open your writers to creating duplicate entries. Many CMS tools will have a Glossary type of object, with sorting built-in. Try the right-click menu if you don't see this feature immediately.

Note about delimiters: Some browsers and PDF readers ignore punctuation when a word is searched for. Some do not. Some ignore punctuation in English and do not ignore the character in non-Latin translations. I find it best if the delimiter leaves white space after the term. A dash (with a space before and after) or tab is better than a colon or dot. For example "term:" might not be hit by a search, while "term" is hit.

6.8.2. A Rose is a Rose in All Our Guides

All definitions of a word must be identical. If each writer creates their own definitions, you might create conflicts between your deliverables. I've seen two guides from one release, with the same term written differently and conflicting each other: "X is Y and is installed on Z" and "X is Y and replaces Z".

Create a master Glossary. When a writer defines a term, it is added to the Glossary. When a writer adds a Glossary to a deliverable, they use only the master terms and definitions. This saves a lot of time, reinforces consistency in writing, and shows consistency and stability in the organization's products.

6.8.3. If It Don't Mean Much, Don't Doc it

The deliverables should not include the complete master Glossary. Each deliverable should include only the terms mentioned in that deliverable. As a proofreading task, you can search for each word in the deliverable's Glossary. If it isn't used, remove it.

One way to make sure that important terms are used, is to run an instance-counting script on a completed document.

6.8.4. Glossary as a Marketing Tool

If the Glossary is in the beginning of the book, it can be a marketing tool. It can plug new technologies and features. If a user opens your guide thinking, "I wonder if this release finally adds anti-bot features," and *Anti-Bot* is the first word in your Glossary, the user is going to smile.

I'm not saying that Marketing should write the Glossary. But letting them have a say about what terms to use or remove might not be a bad idea. Make sure that you do not use terms that are against the image of the organization. If the product is for computer security, use *Anti-Virus* and not *Virus*, and make sure *Hacking* is not a term in the Glossary. If you are writing a layman's guide to flying a drone, do not have a term for *Spool* with a technical definition that only a helicopter pilot recognizes as *increase the blade speed*.

6.8.5. Consistency is Key Again

Add a section in the Style Guide for writing Glossary items. These are some questions to consider:

If the term has an acronym or abbreviation, should the short form be the term or the long one? Pick long or short, and enforce that. Then decide how the other form is given. For example, if the short form is the term, is the long form the first or last part of the definition? Is it in parentheses, or is it offset with a dot?

Should the term be a variable, if possible? This helps if names of products and technologies change often, but can ruin automatic alphabetical sort.

To be useful, definitions should be short. Do you want to set a max word length? I prefer to say that a definition must not contain a word from the term. If you must repeat the term (or part of it), you are not defining it. On the other hand, a definition must not use words that are more complex than the term itself.

Does every term or no term have an example?

What is the style of the term and of the definition?

Do you use an indefinite article (*a, an*) always or never for non-group nouns? Do you use the definite article (*the*) for group or plural nouns? (By the way, my personal preference is never to the definite and always for indefinite.)

If the term is a verb, is it always or never the infinite form, with *To*?

Can definitions have more than one paragraph? Are bullets allowed? If there is more than one use for the term (it means different things for different products or audiences), do you use dictionary style (number the definitions in one paragraph)?

Can the definition include instructions for usages or installation? (I prefer not, but it is up to you. Some technical definitions can be meaningless if you cannot talk about how the thing is used.)

Can the definition have links to other places in the document? Be careful with this one. It might seem like a great idea to link to the first usage of the term, but if the link makes the definition longer, you've lost any advantage.

Can the definition have a non-inclusive series (*and so on* or *etc.*)? Personally, I think the non-inclusive series should never be used in technical documentation, but again, that's a personal style.

Can the term have a *See* link? Can the definition have *See Also* links? If yes, what style and line formatting is allowed?

7. MANAGING TECHNICAL WRITERS

After a lot of thought and moving things around and back again, this section ended up after *Managing Documentation Projects*. I think the first section is useful for all technical writers and managers. This section is more useful for managers than solo writers.

7.1. Building Your Team

The questions most often asked of me from new managers who were not writers were: *What do I look for in writers? How do I test them? What can I expect from them?*

7.1.1. Desired Skills

The line between a good tech writer and a great tech writer is not fine. It is broad and easy to define. These are skills you can expect of an experienced tech writer.

Skills of Good Tech Writers	Skills of Greatness
Knows how to apply styles and diligently paginates PDFs and printables for best formatting	Knows how to create macros or scripts to style and paginate automatically, while diligently verifying automated output
Knows HTML 4.0	Knows HTML5, CSS3, RWD, and keeps up with web technology without impacting schedules
Is professionally fulfilled most days	Is personally fulfilled by a conscious awareness of how the profession helps human progress - though doesn't talk about this much, not wishing to appear to be a crackpot
Writes with correct spelling and grammar	Understands the linguistics of the source language, writes correctly, and can explain why it is correct and when exceptions are acceptable
Is a detail-oriented perfectionist	Is a detail-oriented perfectionist who prioritizes tasks to produce "good enough" on time and can live with the results

Skills of Good Tech Writers	Skills of Greatness
Communicates well, verbally and in writing	Communicates well in all forms, with a predominantly positive energy and an innate understanding of appropriate tone, tact, and timing (In other words: does not piss off contacts, does not write novel-length emails, and can present to a larger group.)
Implements change requests quickly	Implements change requests quickly and thoroughly, diligently verifies that the deliverable contains all changes before sending it off, never making excuses but explains briefly why a change cannot or should not be done

7.1.2. Desired Characteristics

You can find competent writers who do not have these characteristics, but they are in every great tech writer I ever met. This table shows characters to discover in the interview process, why they are important, and how to discover if the applicant has them.

	Importance	Discover
Self-learner	Writers can burn SMEs with questions that can be answered with research.	Ask if they prefer to learn from self-study or from interviewing SMEs.
Quick learner	The more quickly a writer learns the technology, the more time for other tasks and the faster you can tell people to RTFM.	Look for logical associations between existing knowledge and new concepts.
	Be careful. A nervous writer leaps to assumptions that short-cut the learning experience. These assumptions cause mistakes that reviewers do not catch. The keywords are all there. Everything looks fine until someone tries to actually use the documentation.	

	Importance	Discover
Tech lover	Productive tech writers are interested and challenged in the daily work. They love learning about new products and about new writing tools and innovations.	Do not be fooled by applicants who claim they study coding on their own time. Look for applicants who get excited about the tech they used or created (such as macros, scripts, or VMs) as part of their work.
Focused	Technical writing is detail-oriented, sometimes tedious, usually multi-faceted. Focus is necessary to get a final deliverable on time.	Ask for their multi-tasking strategies. Their answer is usually a good lead-in to questions about focus.
	Note from my experience: Active parents are focused employees. They have their end time in front of them, every minute of every task. They can't be late to pick up their kids. If parents can work in the hours of their SMEs' office or virtual availability, you are golden.	

Now you have a post that describes your expectations and desired skill set. The next steps are interviews, making a choice, and onboarding.

 I hate the word *onboarding*. The creation of a noun from a phrasal verb...ugh. But sometimes we just have to go with the flow.

Coaching Desired Characteristics

Before we move on with the next steps, let's take a look at the characteristics again. During the interview and test, make notes of indicators that an applicant might require guidance. Consider applying some of these solutions as preventative measures. Keep them in mind during your new employee's first three months.

Characteristic	Coaching
Self-learner	Make a procedure that no SME is approached until the writer learned the basics. Make sure writers have access to research and non-human resources. If an SME complains of a writer's lack of basic knowledge, ask the writer to work through the issue: What knowledge was I missing? Where can I learn it on my own? Do I need to take an online class? Do I need practice? Can I create virtual machines? What can I do next time to improve?
Quick learner	Find the cause for mistaken assumptions. If the writer is too anxious to prove skills, make changes to the schedule and goals to give the writer more small wins. If the cause is ignorance, ask the writer to write an issue report, to find for themselves the mistake, its cause, and how to prevent it.
Tech lover	Understand why the product is important and convey the awesomeness to your writers.
Focused	Some employees are so enthusiastic about learning, they cannot get their work done on time. Help them focus with: • Task priorities and deadlines • Clear objectives and expectations • Tracking and analysis • Scheduled time for the tasks they love • Defined responsibilities to be the team expert of relevant technologies (with scheduled hours to indulge in research)

7.1.3. Interviewing Writers

The choice of the best applicant feels almost random. Some people are excellent at tests and interviews, but not-so-much when it comes to doing the actual job. And the process of interviews and screening drains our time and often our emotions. Here are some tips to make it as effective and easy as possible.

If at any point it becomes clear the applicant is not right for you, stop. The applicant may protest when the interview ends early, but stick to your gut reaction. It is a kindness to not waste their time or yours. What would make someone unsuitable? Those who require facilities you cannot provide (private office, extra screens, new noise-canceling headphones), situations that are against company policies (continue working a second job or freelancing while working for you, demand to work fewer hours), or authority beyond the position (demand to use their own style guide or tools).

The first time I ended the interview after ten minutes was a liberating experience.

Preparing for the Interview

Send the applicant an email or text message. Ask for a convenient time to call.

I dislike making an unexpected first call. I hate not having my calls answered. And I detest speaking with someone while they are driving. If my call were to ever be the cause of an accident, I would never forgive myself.

1. Make the call according to the applicant's time. Plan for 10 to 15 minutes.

    • Ask one main question about a skill that is important to you.

    • If the person seems overqualified, ask if they are sure they want this position. Explain the job requirements in more detail.

    • If they do not have a keyword on their CV, ask what they know about that tool or technology.

For example, when I was looking for a tech writer who knew Responsive Web Design, I escaped five unnecessary interviews with: "Your application says you know HTML, but do you know HTML5 and CSS3?"

In each case, the applicant said they could learn it quickly because HTML is easy. Personally, I learned HTML in one weekend, back in the 1990s, but when I started an online course in Bootstrap for CSS3, three weekends were not enough. People, please! If I say I want HTML5, CSS3, and RWD, do not toy with me!

2. If the call goes well, schedule an interview time.
Tell the applicant you will send an email with all the details. Have your schedule open while you set it up with the applicant. Make sure the time is free before and after. You need time to review the CV, prepare a form for notes, and maybe prepare a laptop for tests.

Make sure the hour is during your normal working hours. If the applicant can only come before or after work, it is best to make it after work.

 Twice I forgot that I scheduled an early interview for a day that I don't usually come in early. I still blush about being late. Job seekers spend time and money getting to interviews. It is more than rude to waste their time – it can be damaging.

If the call does not go well, be kind. It's a small world.

3. Immediately after a successful call, send a form email.
(The first time you write this email, save it as a template or Gmail canned response.)

- Change the greeting for this applicant's name.

- Start with: "We scheduled an interview on *day*, *date*, at *time*."

- Give directions on how to get there and how to park.

- Tell what to bring (For example, if an ID is required for security).

- How you want samples (Email before the interview? Printed?).
Many applicants bring a USB. Ask your Security Officer or IT if this is allowed. Most organizations have Device Control security against random USB injection.

- Your phone number to call if there is a change in schedule, they get lost, will be late, or have questions.

- Make sure your name is on the email. The applicant only heard it once during the call. If they forget or did not hear you well, it can affect their confidence. You do not want to inadvertently put more pressure on applicants. *Pressure* is something to save for a purpose.

- Tell how long you expect the interview to take. The first one-on-one should take about 30 minutes. If you plan to give a test, add ten minutes for a break to the test time. If you expect the candidate to meet the next interviewer on the same day, add a second hour. Do not explain what the time is for because you may decide not to give the test that day, or the second interviewer might not be available.

4. Make a test.

   Without a test, you do not have an objective stance when a higher-level interviewer asks why you want that person. Test results will also show who is better suited for your team when you have a few applicants. When I did not give a test, I regretted it. Let the applicant decide if they want to take it after the interview or on another day.

Interview Form Template

Interview Form	Print out one for each applicant. Keep notes (this form) with CV and test results	Name: Date: Contact info:
Question asked in call:	Answer rating =	[ ] Sent email with directions
Samples:	English Quality =	/ 10
	Formatting =	/ 10
	Content =	/ 10
Open Discussion:	Interesting notes:	
	Good questions:	
	Communication skills:	
Leading questions:	Salary:	
	Self learner:	
	Experience:	
Test:	Follows instructions	/ 10
[ ] Same day?	Shows flexibility	/ 10
[ ] Desired environment	English quality	/ 10
	Time (exact)	/ 10
[ ] Qs on the side	Independent	/ 10
[ ] 60 min timer	Technical	/ 10
[ ] Escorted (not alone)	Added value	/ 10
References	Name / Contact / Relationship	Summary
	Name / Contact / Relationship	Summary

Interview Form Example

EXAMPLE	Name: Joe Average; Date: 4 May 2018; Contact: 555-123456	
Q asked in call: Asked about CL experience	Answer rating = 5 (had a clue)	[v] Sent email with directions
Samples:	English Quality = 2 grammar mistakes	5 / 10
	Formatting = fine, a few style calls	5 / 10
	Content = too much marketing	5 / 10
Open Discussion:	Interesting notes: EXP SHAKY. Likes tech & writing, no long work place	
	Good questions: FLEXIBLE. Asked about style guide, has used many, written own	
	Communication skills: SO-SO. To the point, didn't understand all Qs, maybe nervous	
Leading questions:	Salary:	
	Self learner: UNSURE. Said he is but talked a lot about interviewing SMEs	
	Experience: 10yrs. SAMPLE NOT GOOD ENOUGH	
Test:	Follows instructions	5 / 10
[V] Same day?	Shows flexibility	5 / 10
[V] Desired environment	English quality	5 / 10
	Time (exact)	5 / 10
[V] Qs on the side	Independent	5 / 10
[V] 60 min timer	Technical	5 / 10
[V] Escorted (not alone)	Added value	5 / 10
References	Alef N. 555-12345. Frmr boss	Left when it got hard, avg productivity

Making the Applicant Test

This is not the most pleasant way to spend your time. I wish I could be more help, but I cannot give the best test. It must be different for each company. I can give you some ideas from my own lessons learned.

Creating Test Content

- Do not give an editing test to a writer. Do not give a writing test to an editor.

- Test for assumed knowledge.
  If you expect all your writers to know the lifecycle of a document, put that in the first part of the test. It can be an open question, or a numbered list with blanks to fill in.

- Make the test similar to daily work.
  If your writers create procedures from technical emails, your test can be a long email, from which the applicant creates a procedure. If you are concerned about proper formatting, include this in the test instructions. If your tool is XML-based, the test instructions can say that formatting is not important. If an applicant wastes too much time on formatting when it is not important, this is a good indicator of problems with flexibility or following instructions.

- Give the same content to all applicants for this particular position.
  Watch for applicants who know too much about the test. Once, I tested an applicant who knew the test contents ahead of time. Our test had been leaked to a head hunter. When the test is compromised, you will have to make a new one. If this happens before the current position is filled, make the new test as similar as possible.

- Have an answer sheet.
  Give the test to your team before you give it to applicants. This will give you a benchmark for time and you can merge their tests to have the best answer sheet. This control also sometimes indicates that veteran employees need a refresher on the style guide or methods. That can tell you who would gain the most by mentoring a new hire.

Giving the Test

1. Before the test, tell the applicant how much time they will have. Ask if they need anything before they begin, such as a glass of water or a restroom break.

2. Do your best to put the applicant at ease. Some people get nervous at just the mention of the word *test*.

3. Try to make the test environment fit the characteristics you are looking for. If you want someone who can focus in a high-energy, noisy environment, avoid giving the test in a silent bomb shelter.

4. When the writer is seated, give instructions.

    It is best if you say, "During this test, if you have questions, you can write them on the side." If you offer to be the SME for technical questions, you open yourself to giving each applicant a different amount of help. Avoid interacting with the applicant during the test. If you must help an applicant during the test, make a note of that fact. If the applicant is entry-level or an intern, asking for help might be a good sign. If the applicant comes with a 2-page CV of tech writing experience, and still asks basic questions, you'll want to remember that. Or you might want to end the interview there.

5. Make the test timed.

    When the writer is ready to begin, start the timer. Go ahead and put the pressure on, especially if the work will be high-pressured. You do not want someone who thinks, "I love deadlines. I love the whooshing noise they make as they go by."[7]

6. During the test, do not leave the writer alone unless you must. Watch their reactions and body language as they work.

7. When the test is done, if you look it over in front of the applicant, do that for all applicants. This can be useful to see how the applicants take criticism. But you do not want to get in a situation where they ask, "How did I do?" and you don't want to tell them. Say that you will get back to them after seeing all applicants.

Giving an Effective Interview

Do not ask interview-basic questions. Start with, "Tell me about yourself, as a technical writer." See where that goes. Follow up with questions, like a real human conversation. It's more fun that way.

If the applicant talks too much during the job interview, they might also talk too much during SME interviews. Ask about this trait when you call the references.

Of course, you want to lead the discussion, to learn if the applicant fits the characteristics you are looking for. For example, if you want someone who will do self-study before they approach an SME, you could say: "That sounds like you were writing about a ground-breaking tech. How did you go about learning enough to write it?" You can be direct if you do not get an answer to the hidden question. "Do you prefer to learn on your own, or to ask others?"

---

[7]Douglas Adams, *The Salmon of Doubt: Hitchhiking the Galaxy One Last Time* (UK: William Heinemann Ltd., US: Pocket Books, 2002).

Be true to your priorities. I once took a writer who answered that last question with, "It's always fastest to ask." His test was excellent, but on the job he spent more than 25% of his time in hour-long meetings, usually misunderstanding what the SME was trying to explain for the tenth time.

(I changed the gender of the applicant to protect her...or did I?)

During the interview, ask about something from the samples you do not like. Watch how the writer accepts criticism:

- "I didn't have a lot of time with that one." -- If the applicant cannot accept writing critiques without taking them personally or making excuses, it is a sign of professional immaturity.

- "This book was written by many people. Only some of it is mine." -- Why did you try to pass it off as a sample of your work? Look for signs of unethical behavior.

- "I wrote that years ago." -- Why didn't you bring your best? Look for signs of professional inertia, a lack of urgency to get a job and do it well.

I am still not sure about the "What do you know about our company?" question. I like to know if an applicant did their homework, and if they have a passion for the industry. When that question was on my interview to-do list, it started me off with long explanations of why I like the place I'm at, what we do on the team, and what products or services the company delivers. Truly, I wish someone had duct-taped my mouth sometimes.

Remember to ask: "Do you have any questions?" When I stopped talking about the company without invitation, applicants usually asked about it. They always ask about pay. And the best answer is, "I don't know about the pay of my employees. That is confidential. Do you have an idea of what you are looking for?"

Of course, this is the best answer only if it is true. I never knew the salary of my employees, and I never wanted to know. I want bonus and raise suggestions to come from performance, not comparisons or need.

If the applicant has a strict minimum, that is worth noting. If it is too high for the company, you saved a lot of time for the next interviewers and the applicant. If they are not sure, do not push for a number and do not write it down. If they see you writing, they will think that you gave a rough estimate to HR and might see a low offer as an attempt to gauge them.

After the Interview

- Check references.

    A person's track record is much more important than anything said or done in the interview. I'm not a phone person. I don't like verbal conversations with people I don't know. But you must call those references.

- Look at the applicant's online profile.

    I try to avoid non-professional pages. I do not want someone's political or religious views to sway my judgment. I do look at LinkedIn references. If a reference is in their LinkedIn network, that's awesome. I can learn about the reference before I call or send a message before I call. That makes me more comfortable. Sometimes that leads to having the whole conversation in text, which saves me a lot of time.

    If the person listed as a reference has a referral from the applicant on their own LinkedIn (a quid-pro-quo friend), I am wary. I assume I am not going to get an objective reference.

- Follow up quickly:

    - For applicants you want to hire, push the next steps to be done as quickly as possible. I've lost more than one applicant because they found a job before we were ready to make an offer.

    - For second-choice applicants, do not send a rejection letter until a candidate signs on. It's not over until the fat lady sings - or the applicant signs. And you do not want to lose a good applicant because you rejected too quickly. Send a message that the process is ongoing and to please be patient. Let them know when the process is done.

    - For rejected applicants, send a form letter. Be kind. In your records, note that you communicated the termination of the process. Do not throw away CVs or notes about rejected applicants. They might apply again later, or they might be sent to you from a colleague or head-hunter. At that time, they might fit better. If not, you will want to remember why you rejected them. Or you might want to call the applicant when a new post opens. One of my favorite employees was a call-back.

    - If a rejected applicant asks why, do not respond, or do not respond with a direct answer. I made this mistake once. The quick email became a thread and then became quite unpleasant. Consult with HR. They might have a policy about this.

7.1.4. The First Day

When your new hire arrives on the first day, it should be like clockwork. Ask HR about the procedures. What day and time will the candidate start? Do they sign paperwork first? Do you meet them in the reception area or can they come to the team's offices immediately?

If there are no standard procedures, write them and offer them to HR for review and approval.

Preparing for the First Day

- Assign someone on the team to be a mentor to the new hire.
It is best if you can do this as soon as the applicant signs, before they actually start. This gives the mentor time to consider what they think would be helpful and to bring schedule conflicts to you.

- Prepare materials.
If the company does not always have a workstation ready on the first day, have materials for the new hire to read. These can be information about the products, your team Style Guide, your standard of the Controlled Language you use, and books that you would like your writers to read.

I usually loan my copy of C. P. Snow's *The Two Cultures*. I used to also hand them my **Strunk and White**, until it went missing. *grrr*

- If you have a handbook with the company core values and team procedures, give the new employee a copy.
We had a handbook on the internal wiki, but I also printed it once, to share around. New employees were encouraged to write notes and corrections in it. Many were more comfortable making changes to the wiki only after they had approval for their changes in writing.

- Create a training schedule, and put it on an accessible network, such as your team's wiki space.
The writer should know exactly what to do now and what to do next. Your training schedule will have links to resources, estimated time to complete each module, and measurements for successful completion of a training module.

Training Schedule Example

Area	Resource	Measure Success	Due
Product	In-house lectures	Write product notes and knowledge in wiki & approval by TL	Availability
	Existing Docs	Find errors in docs – at least one technical	1st Wk
	Industry Docs	Create one use case and one feature request	1st Mo
Writing	Style Guide	Q&A with mentor, continued usage in writing	1st Wk
	CL Standard	Q&A with mentor/TL, participate in CL forums	1st Mo
	Team wiki	Own pages for projects, doc issues & status as 2nd writer	1st Q
	Tool tour	Demo tool usage in 1st project with mentor shadowing	2nd Wk
Advanced	Macros	Sandbox and proper procedures to automate discovered issue	1st Yr
	Scripting	Automate one task or create a new team feature with a script	2nd Yr
	Networks	Q&A with TL	2nd Q
Soft Skills	In-house training	Time management, Presentation skills, Meeting skills, Other skills on request and approval of TL	Availability (after 1st Mo)

First Day Checklist

☐ Mentor _____ confirmed hours scheduled for mentoring (Mentor and manager agree the hours will be used as the new employee needs them. This commitment blocks off time for mentoring, to re-arrange projects.)

☐ Training schedule and materials are available.

☐ Workstation is available. If not, due date: _____ and new hire goals until
then:_____

☐ Lunch with manager scheduled.

☐ Company policy is available.

☐ New hire proved understanding of company policy and correct behavior.

☐ End of first-day meeting scheduled.

☐ End of day meetings scheduled for first week.

☐ Weekly meetings scheduled for first month.

During the First Day

Introduce your new hire to the team and employees in the area. I think it is best if you also clear your schedule to have lunch with the new employee. The first social interactions are very important. If you miss out on it, the new employee's feelings of being overwhelmed and lost can last much longer than usual.

When the writer has a workstation, the first thing to do is often set by company policy. If your company asks employees to watch behavior videos or read through security practices, make sure that is done.

Then, the new hire begins the training schedule.

Schedule a meeting for an hour before you or the new employee ends the day, whoever leaves first. You want to know how they feel, ease their worries, and find out if they need more help.

Questions to ask at the end of the first day:

• Is your workstation comfortable? Do you need or want anything to make it better?

Touch-typing writers should have ergonomic keyboards. You do not want to lose a person to carpal tunnel syndrome.

- Can you explain what you did today and the purpose of the tasks?

- What are your goals for tomorrow?

- Do you understand the company policies and core values? How would you explain them in terms of your daily work as a technical writer?

Do not settle for "Yes." Make sure you know the policies and values and that you have an answer for this question.

- Who on the team helped you?

If it is not the mentor you chose, have a conversation with the mentor the next morning. Maybe the mentor has time concerns that you must handle with scheduling.

- Do you have questions?

- Can you give me feedback to improve onboarding procedures?

Use the questions to get to know your employee better. Do not take feedback personally. Do not make excuses. Do not judge. If your new hire does not have anything to ask or offer, suggest that you will be open to feedback in the coming days.

During the first week, schedule similar meetings according to the person's needs. If your desk is not near theirs, they might feel disconnected and need more face-time. If you can, stop by once or twice a day to see how things are going. Without creating an atmosphere of micro-management, see if the new employee is comfortable or needs more help.

At the end of the week, review your expectations and how they are measured with the new employee. You want everyone to come in on the second week, and every week after, with: "I know what to do, and I know how to do it."

Company Policy Implementation Example

A quality of large, stable organizations is the existence of understood realistic values that drive employee decisions. If your company has a policy or core value system, make sure you can interpret it for your daily work and the behavior of your employees. If your organization does not have this set up, you can make core values for your team. Here is an example.

INTEGRITY	Do not install illegal software
	Be honest in your work and in your dealings with everyone
	Do not speak about unreleased products outside of the company
	Respect the resources – printers, paper, computers, bandwidth – as though they were your own
	Do not use your office time or other resources to work on non-company projects
	If you make a mistake, be responsible for the solution
CUSTOMER FIRST	Make sure you give the customer what is necessary to succeed
	In any question of the customer needs or sales needs, choose the customer
INNOVATION	Welcome changes
	When you see a way to improve yourself, the team, or the company, drive the change through the appropriate procedures and hierarchy

7.1.5. Expanding Your Team

You must know, and be able to justify to your management, the number of people you need.

Every successful company spends carefully on salaries, letting overhead decrease net profits as little as possible. Then the company reaches the threshold. They realize (hopefully in time) that they will lose deals if they do not provide updates, products, or deliverables. The company looks at the three points of project management: time, budget, resources.

If time is an issue for a deal, and there are not enough people to make it happen, they will spend the money to get more. You can get a new hire, if you can show there is enough work for more permanent people. If you cannot show this, you will get temps or contractors. That is not an inherently bad thing, but it does cost time to train new people - a cost that is not regained when the temporary contract ends.

If you are always putting out fires for an overworked team, you will not be able to prove that you need more people. Feel free to reject projects, based on priority scheduling. Become trustworthy with estimates (I'll show you how later) and wave that red flag from the start and every day, when you know your team cannot meet the deadline. And then don't meet it. This is really important. Your team must fail for it to grow. Let me put that in its own space:

If you do not have the required resources, **your team must fail for it to grow**.

If your team grows without you saying how much additional headcount you need to meet the next quarterly and yearly requirements, start looking for another job. Your company is wasting budget and will be bought or go under. BUT (and this is a big one), if you can warn of "failure" before it happens, you can get headcount approval in time and grow in a natural and successful progression.

If you can prove the need for more headcount in time, and still cannot get approval, make sure your team succeeds on priority projects. Protect your team from unplanned issues, as much as possible. Spend your time on management: track work, create estimates, set priorities, meet with your team and SMEs, and track rejected work. Do not work overtime, and do not ask your team to work overtime. You planned properly, to get the work done on time with the necessary man hours. If you put in overtime, your management will not understand the need to hire more people. From their perspective, there is no gap to fill.

7.1.6. Letting Go

If you think you must fire an employee, that might be the correct thing to do. Before you go there, consider your attitude to the person. Were you fair? Did you let that person have a say often enough? Did you consider illness or external situations, such as home life? Does that person trust you enough to confide in you about such things? Did you weigh the time that the employee costs you against the time necessary to find and train a new employee?

Consider the Rotary Four Way Test.[8]

Of the things we think, say or do:

1. Is it the TRUTH?

---

[8]Rotary International. https://my.rotary.org/en/guiding-principles

2.   Is it FAIR to all concerned?

3.   Will it build GOODWILL and BETTER FRIENDSHIPS?

4.   Will it be BENEFICIAL to all concerned?

If you have nothing to improve in your behavior in this case, and you made a fair effort to improve the employee's behavior, consider going to HR. First, you should not even think of firing a company employee without consulting Human Resources. According to the protocol of the organization, you might not have the authority to fire anyone. Second, if you have never consulted with your HR, you are missing out. This is their area of expertise. You might be given a great new idea that will solve the issue and make you realize you should have come to them sooner.

If you do not know or trust your HR, think of this: one of the best ways to make a friend is to ask for advice.[9] Having HR on your side will bring you a wealth of expertise, sound advice, and ease in new endeavors.

If you are sure you must fire an employee, know this: it is the most difficult task a manager can face. Your methods might become more fluid with experience, but the emotional impact does not decrease. If you find yourself firing left and right without personal impact to yourself, you don't know your people as well as you should.

Make sure you know the legal ramifications. Make sure you notify everyone in the organization who needs to know ahead of time. And follow through.

Make the process as easy for the employee as possible, but stick to your goal. Be objective and have real reasons ready for the typical "Why????" or "Yes, but," that will come. Best: you warned the employee that low productivity, bad attitude, or whatever it is, is an important issue, months before you got to this point. The easiest change in employment status comes when all parties are aware of the probable outcome long before it actually comes.

---

[9] Appel, Allen. From Father to Son: Wisdom for the Next Generation. New York: St Martin's Griffin, 2017.

Probation - Termination Checklist

☐   Employee understands expectations

☐   Employee agreed the expectations were realistic and could meet them
OR: employee expressed doubt about _____

☐   Expectations are fair and equal for all members of the team

☐   Expectations are beneficial to the company

☐   Employee communicated these issues:_____

☐   Solutions to employee issues were planned and agreed on
OR: solutions not found due to: _____

☐   HR and Managers support termination

☐   Time to train newbie is insignificant compared to issues of current employee

7.2. Working as a Team

For some reason, tech writers think they a species apart from QA engineers and R&D developers. (Then the writers complain about being excluded.) The truth is, while models for development and testing teamwork were born, grown, flown, or died, many companies leave their tech writers alone. Many tech writers do not have experience working with other writers as a team. Testers and coders have long since learned to break up and share tasks, but not many tech writing teams have yet bought into the scheme.

Let's look at this scenario: Alex is working on Project A. Betty is working on Project B. Project B hits a snag, and developers estimate a month's delay to fix the bugs. Betty has some unexpected time on her hands. Project A, on the other hand, is moving along nicely and is going to be released early for a Proof of Concept (POC) customer. If Alex tries to handle Project A by himself, he will have to work 2 hours more every day until the early release.

If Alex and Betty were QA engineers, Betty could easily pick up test plans from Alex, and the team would complete their work on time.

If Alex and Betty are tech writers on a team that works as a team, Betty will easily pick up doc tasks from Alex. Your team will complete both projects on time, without overtime.

Let's make the scenario of Alex and Betty more complex. Alex is called up for reserve duty. Now you have no writer for priority Project A, and a full-time writer for the currently on-hold Project B. You open Alex's project plan (because you work as a team, the plans are made, stored, and updated in a path that everyone can access). You scope the remaining tasks of Project A. With Betty, you set expectations for her to complete the tasks in time. You communicate the writing stop to the manager of Project B, for X days, and you tell the manager of Project A that Betty is on it.

You must tell the Project B manager that Betty will be unavailable. While development is focused on specific fixes, the manager has time for the documents. If you do not talk about the change in schedule, Betty will be constantly pulled in different directions.

Betty opens the project plan that Alex started. It is in the same template as her plans. She picks up Project A easily. And the flow continues, on time, with the same quality, without higher costs.

Steps to lead the miracle team:

1. Recognize that tech writing can be a team effort, just like coding or testing.

2. Communicate that to your team.

3. Build and enforce a foundation for consistency in the results of all team members.

4. Use a tool that enables teamwork.

5. Use a metadata system for efficient handover.

7.2.1. Step 1: Recognizing the Possibility

For the first step, yes, tech writers can work as a team. Some organizations already do this. Multiple technical writers join in doc sprints, handle project replacement, communicate the impact of handover to lower priority project stakeholders, and complete priority projects as a team. They pick up for sick and absent teammates. They share workloads.

Teams of writers around the world deliver technical communication in all forms, on time and budget, at quality benchmarks, with teamwork on shared projects. In the 1990s and early 2000s, you could get away with one writer on one product. But that doesn't fly with Agile development, sprint Releases, and the evolution of technology development. You and your team must be dynamic. Recognize.

7.2.2. Step 2: Communicating the Change

For the second step, you must have a team meeting schedule and trustworthy estimates. To make the flow to teamwork natural, you can start when it is required.

For example, if Alex knows he will have reserve duty next week, you ask him to do a knowledge transfer with Betty. She goes over the plan with Alex, but she thinks that she cannot meet his estimated dates. She does not have his previous knowledge. When she comes to you, you suggest that Betty be the lead writer and that Charlie handle the one feature that will cost Betty more than two days to complete (where Alex or Charlie can complete it in two hours). You schedule an urgent meeting (made for the same week, no exceptions - everyone will have to juggle their other meetings if there are conflicts) to handle the issue that Project A needs multiple writers.

To prepare for the meeting:

1. Learn from R&D how much time until the early release milestone of Project A. Find someone who will give an honest and realistic answer.

2. Get a summary from the writer.
   The summary is in a project plan and includes: features written, features not written, proofreading status, SME approval status, screenshots done and screenshots to change

or to create, graphics (done, to change, to make), all SME names and commitments to review by date, committed deliverables, and details of how to run hands-on. (We will talk about the project plan later.)

3.   Of the remaining work, get estimates of required effort (in hour or day-fractions).
     If the writer cannot give estimates, and you do not have past tracking to guide you, lead the estimates. Do not get this information in a verbal communication. Writers like to talk and work things out in their minds as they spew their confusion and frustration all over their listeners. Get this in writing. If the writer is experienced, send an email after your project review meeting with a due date. If the writer misses the due date, or is not an experienced planner, write it out with them.

4.   Bring in the second writer, if you already know who it is. Ask for their opinion on the estimates.
     Can the second writer meet the estimates? Does she have the technical background to do it faster? Or is she unfamiliar with the technology and unable to commit to the expectations? Update the plan to make sure the writers can succeed.

5.   Before the meeting time, get supporters for the change.
     This is an important step in Change Management. Do not skip it! Find the person on your team who has positive energy and a positive outlook to change. Tell them about your plan. Ask for advice on how best to reach your goals with all the individuals on your team. You can bring in your manager too, but only when you have all the answers to "What happens with the other projects?" I would do this only if your manager also has a positive outlook, especially for innovation.

6.   Prepare answers for predictable questions and fears.

7.   Schedule the meeting for the whole team.

Of course, you can meet with the priority project writers for a stand-up status every day, but that is different. This is you, turning a problem into an opportunity for everyone. You will work as a team from now on, and continue to reap the benefits in the future.

In the team meeting, tell your team (in whatever way you decided was best):

"We are going to work as a team. Project A needs X writers to complete on time."

If you cannot answer all questions in the meeting, schedule a follow-up brainstorming meeting. Every question has an answer, but some are not immediate.

7.2.3. Step 3: Enforcing Your Ese with a Style Guide

You know that Time magazine has *Time-ese*. Show me an article from National Geographic, and I bet I can recognize its *Natgeo-ese* in ten words or less.

If you have an *ese* for your corporation, you will have consistency, no matter who actually writes the words. You create your content consistency - your *ese* - with an enforced style guide. This is a big deal, so it has its own section.

Personally, I think that you also must have a controlled language (CL), but that's just my experience talking.

CL is also a big deal and has its own section.

7.2.4. Step 4: Choosing a Tool for Teamwork

For the fourth step of working as a team, you need a tool that allows for collaboration. Many tools can be tweaked or integrated with other tools to allow collaboration. If you are stuck with a single-writer tool, you might have to choose a tweaked or integrated solution.

For example, if you must use MS Office, enforce pathname conventions on all files and use Share Points. You can add VBA macros for collaboration features, such as the status of sections. Use Word features for re-use of sections in different documents. Look at tools that integrate Word files in a Content Management System.

If you have a budget and can choose your own tools, look for a Content Management System (CMS). Better, if you have a budget and a choice, and your team is flexible enough, look for a Component Content Management System (CCMS). The difference is a CMS manages deliverables of all types, with the writing in files that can be complete documents. They are not chunked. A CCMS is designed for re-use, with the writing in topics or chunks.

7.2.5. Step 5: Covering Your Assets with Metadata

This is extremely important. A team is effective as a team only if each person has the ability and skills to pick up the work of every other person. Where is the source container for every project in process? How does every person on the team, anywhere in the world, get to the source of everyone else? How does a second writer know the variable and conditional assignments to use when publishing someone else's book?

Metadata will also save your buttocks. This happens in every company. A writer gets data. A writer refines the data to technical documentation and gets approval. And then a Johnny-come-lately from some other department, after the release of your deliverables as part of a product, asks why that information is incorrect. If you have metadata on the information (date, SME name and quote, questions, verification with hands-on) you can explain why. Your explanation may put an end to the question. Maybe this last reviewer realizes his or her own mistake. Or maybe the information is incorrect. Now the last reviewer can discuss it with the original SME, and you can get the best information and make a change. In both cases, right or wrong, if you have the metadata, you are never at fault, in the sense of what went wrong. Likewise, if you do not have the metadata, you are always at fault. If you do not have the SME name, and you do not have the time to investigate, you will change the doc quickly for the last reviewer. This change will often be incorrect or half-baked. And the fire under you gets hotter every time you touch that doc.

A technical writing team is expected to document their work. If you do not have metadata, your team looks sloppy.

7.2.6. Your Party Begins the Quest

When I was a child, playing Dungeons & Dragons with my siblings, my favorite part of the game was setting off on the quest. Nothing compared to looking around the circle at our fellowship and anticipating the adventure to come.

Documenting as a team is a lot like going on an adventure.

- You are not alone.
  When a monster task beyond your level suddenly appears in your path, you can trust in your party's synergy (skills that are greater than the sum of its parts).

- You have a mission on the horizon.
  You may never bring back the queen's handkerchief from the dark tower (or produce award-winning PDFs), but your journey brings levels of experience that make greater quests possible. Your team will be more capable of producing the next format.

- Decisions are made once and applied efficiently afterward.
  The first time my brother needed about ten minutes to explain "melee-prepared, marching order." It took our party another ten minutes to figure out the order that would best use each member's strengths and protect our weaknesses. We set our marching order before we entered the dungeon. Then as we quested, we simply stated, "melee-prepared, marching order" and in seconds our party was ready. Each member knew, for the individual and for all the members of the party, what to do, when, and how, even before we knew what the task would be. You have your style guide, your plan templates, your tool. You are prepared to go forward.

This team setup is a magical weapon in your quest to reach your vision. You know what everyone is doing. You know the status of all projects. You know when to raise a red flag with your manager and project managers. You know when your team needs help. Writers know enough to be able to jump in the fight when called upon. Projects are given by priority and by people's strengths. A map is kept, leading the team always forward.

To get this magic item, you must visit the armory.

OK, I will end the geek analogy here and explain in documentation manager terms. To reach the stage I just described, you must have these tools: weekly goals, team meetings, tracking, a high-level schedule, and a project plan template.

7.2.7. Weekly Goals

I once had a manager who asked me to send an email the first day of every week. It was based on last week's email, where I entered the status of last week's goals and filled in this week's goals. The effort ended shortly after it began. The emails of last week began to make less and less sense. I couldn't remember all the things I had done, or why tasks were pushed off.

The email system does not work because the plan is static, but our work is dynamic. A technical writer usually does not work on one project every day for a week. We usually work on multiple projects in one day.

A good weekly plan is dynamic. It lets the writer set self-expectations, review it every morning, and update it for milestones, ad-hoc requests, and setbacks. If you try to say, "This is your list. Do it and nothing else," you lose flexibility and burn SMEs.

As Dwight Eisenhower said, "Planning is everything. The plan is nothing."

Weekly plans will change, often and quickly. That does not mean you should not have them. The weekly team plan as a static thing is not worthwhile. The weekly plan as a dynamic container for the act of planning is so worthwhile that it is necessary. The weekly plan lets you and everyone on the team know where the efforts are going. It lets you lead the vision, while they lead their own work.

Weekly Goals or Daily To-Do Lists?

Alex is First Writer on Project X. Betty is second writer for features A, B, C, D, E, F. Both Alex and Betty work on these features. They divide smaller tasks of the features.

The deadline for Project X documentation is in one month. The team keeps weekly goals on MS OneNote.

Alex decides to "make the most of his time." He will do as little over-head work as possible. In the Weekly Goals, this is what they write for the first week:

Alex: Doc X

Betty: X: Learn and write A, B, C, and D. Proofread A-D.

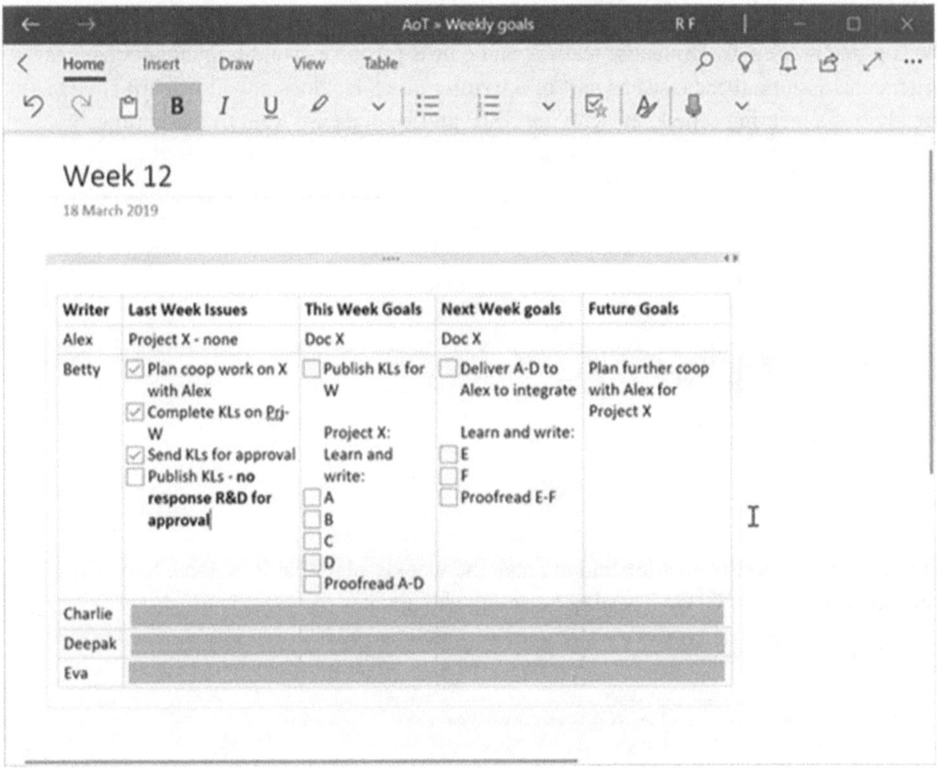

Writer	Last Week Issues	This Week Goals	Next Week goals	Future Goals
Alex	Project X - none	Doc X	Doc X	
Betty	☑ Plan coop work on X with Alex ☑ Complete KLs on Prj-W ☑ Send KLs for approval ☐ Publish KLs - **no response R&D for approval**	☐ Publish KLs for W Project X: Learn and write: ☐ A ☐ B ☐ C ☐ D ☐ Proofread A-D	☐ Deliver A-D to Alex to integrate Learn and write: ☐ E ☐ F ☐ Proofread E-F	Plan further coop with Alex for Project X
Charlie				
Deepak				
Eva				

Week 12
18 March 2019

See the column for **This Week Goals**. You haven't read this chapter yet, so this looks like a fine weekly goal for the team.

Alex begins the work on Monday. He scribbles a To-Do list for himself in his notebook: learn A, write up B, edit C.

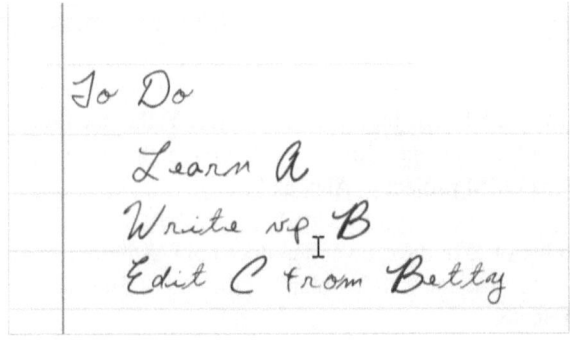

During the day, Alex wrote B and edited C. But he didn't have the time to do A. It was more complicated than he expected.

On Tuesday, Alex's To-Do list is: learn A and edit B (proofread and implement changes - writers can assume these tasks as part of a writing task). He does not add "learn D". He already has a meeting scheduled with the SME and he does not want to waste time.

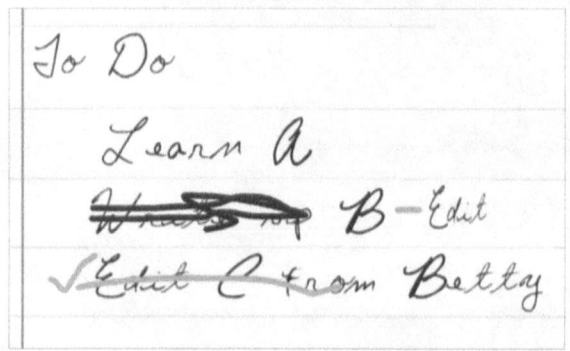

Betty peer reviews B from Alex and updates the weekly plan that B is done. She also completed her part of C. On Tuesday morning, she updates the Weekly Goals.

Writer	Last Week Issues	This Week Goals	Next Week goals	Future Goals
Alex	Project X - none	Doc X	Doc X	
Betty	☑ Plan coop work on X with Alex ☑ Complete KLs on Prj-W ☑ Send KLs for approval ☐ Publish KLs - **no response R&D for approval**	☑ Publish KLs for W Project X: Learn and write: ☐ A ☑ B - peerRev Alex ☑ C ☐ D ☐ Proofread A-D	☐ Deliver A-D to Alex to integrate Learn and write: ☐ E ☐ F ☐ Proofread E-F	Plan further coop with Alex for Project X

Late Tuesday morning, you get a call from a VP. The high-priority Project Y, the document from an OEM provider, must be edited before the end of the week. Everyone on the team is committed to other projects. You give the OEM project to Alex and Betty.

It takes all day. Alex throws his To-Do list of X in the drawer and does not waste time writing down that he is doing Y. Betty adds Project Y to her plan. At the end of Tuesday, Betty marks that the goal for Project Y is done.

Writer	Last Week Issues	This Week Goals	Next Week goals	Future Goals
Alex	Project X - none	Doc X	Doc X	
Betty	☑ Plan coop work on X with Alex ☑ Complete KLs on Prj-W ☑ Send KLs for approval ☐ Publish KLs - **no response R&D for approval**	☑ Publish KLs for W Project X: Learn and write: ☐ A ☑ B -peerRev Alex ☑ C ☐ D ☐ Proofread A-D ☑ Proj Y: Edit 2nd Half	☐ Deliver A-D to Alex to integrate Learn and write: ☐ E ☐ F ☐ Proofread E-F	Plan further coop with Alex for Project X

On Wednesday, Alex and Betty proofread their parts of B and C and peer review for each other. They start to learn A. Betty notes that it is very complex. She sees the plan for the week and knows it cannot be done. She moves the task for proofreading A to next week.

Writer	Last Week Issues	This Week Goals	Next Week goals	Future Goals
Alex	Project X - none	Doc X	Doc X	
Betty	☑ Plan coop work on X with Alex ☑ Complete KLs on Prj-W ☑ Send KLs for approval ☐ Publish KLs - **no response R&D for approval**	☑ Publish KLs for W Project X: Learn and write: ☐ A - underest, complex ☑ B -peerRev Alex ☑ C ☐ D Proofread: ☑ B ☑ C ☐ D ☑ Proj Y: Edit 2nd Half	☐ Proofread A │ ☐ Deliver A-D to Alex to integrate Learn and write: ☐ E ☐ F ☐ Proofread E-F	Plan further coop with Alex for Project X

On Thursday, Alex and Betty complete the research for A. Working together, they complete the write-up of this feature on Friday. Alex crosses it off his list.

Note that Alex did not document that he already learned D. Betty does not know that he has interviewed the SME. If she does not directly ask Alex if he did this, she will call the SME and ask for a meeting. The best-case scenario is a waste of time.

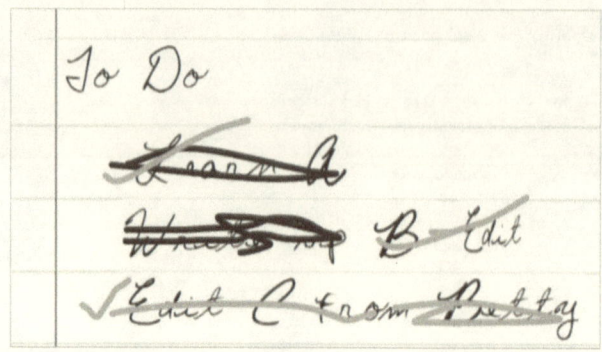

Betty changes her Weekly for the work done.

Writer	Last Week Issues	This Week Goals	Next Week goals	Future Goals
Alex	Project X - none	Doc X	Doc X	
Betty	☑ Plan coop work on X with Alex ☑ Complete KLs on Prj-W ☑ Send KLs for approval ☐ Publish KLs - **no response R&D for approval**	☑ Publish KLs for W Project X: Learn, write: ☑ A - complex ☑ B ☑ C ☐ D ☑ Proofread B, C ☑ Proj Y: Edit 2nd Half	☐ Proofread A ☐ Learn, write, proofread D ☐ Deliver A-D to Alex to integrate Learn and write: ☐ E ☐ F ☐ Proofread E-F	Plan further coop with Alex for Project X

On the first day of the next week, Betty updates the Weekly quickly. She is the first in, so she makes a new page in One Note for the new week. She deletes the column for *Last Week Issues*. She changes the column header of *This Week Goals* to *Last Week Issues*. *Next Week* becomes *This Week*. She sees that A is not proofread and that D must be started. She can explain why they are behind in the original plan: "unplanned high priority project."

Writer	Last Week Issues	This Week Goals	Next Week goals	Future Goals
Alex	Doc X	Doc X		
Betty	☑ Publish KLs for W Project X: Learn, write: ☑ A - complex ☑ B ☑ C ☐ D - unplanned high priority project Y ☑ Proofread B, C ☑ Proj Y: Edit 2nd Half	☐ Proofread A ☐ Learn, write, proofread D ☐ Deliver A-D to Alex to integrate Learn and write: ☐ E ☐ F ☐ Proofread E-F	Plan further coop with Alex for Project X	

That afternoon, Betty leaves at 4 pm to pick up her children. Alex is in the office. Your manager pops in.

"How is Project X going?"

"I'll be over in a couple of minutes to tell you all about it."

...

"Alex, what's the status of Project X?"

"We're done with A to C, but we didn't do D."

"We were supposed to complete up to D this week. What happened?"

"It just took longer than expected."

Hm, that's not an answer to give your manager. You look at the Weekly. Ah! Thank goodness for Betty!

"Hey, Boss! Here's the status of X. A was more complex than estimated, and the OEM project put us back a day. We are ready to deliver B and C. Next week, we will deliver A and D. We also plan to deliver E and F, but we will look again at the complexity of the features."

...

"Alex, please remember that A is not proofread. I think you forgot. During our one-on-one, we are going to talk about your Weekly."

If Alex had left early, we could assume that Alex had completed all of his goals. No other details told us that he didn't meet the first expectations. If you told your manager that Project X was on time, you might end up putting Alex in the position of having to work overtime to catch up.

If Betty had not kept the Weekly goals where you could see them, and all you had was Alex's written to-do list, you would have nothing. You wouldn't know if you were behind or when you could expect to see it completed. You would have forgotten about the OEM project and wouldn't be able to explain its impact on the scheduled tasks.

Weekly Goal Tools

I found success with the free Microsoft tool, OneNote, that comes with Windows 7 and higher. I made a notebook that the team shared.

This team now has a Weekly Goal OneNote set of notebooks for five years!

It had a section for each year and a page for each week in the year.

A weekly page had a table with a row for each writer and columns for: **Last week issues, This week goals, Next week goals, Future goals.**

Every first day of the week, I copied last week's table to this week's page. I deleted the issues column from last week (I didn't lose that information. It was still on the page for last week.) and renamed the next column to **Last week issues.**

Next week became **This week.**

And **Future** stayed the same.

About Future goals: It is very important that your system does not allow for tasks to be dropped without a decision to drop the feature forever. Handle postponed and dropped tasks differently. If an SME says, "There is a critical bug with D. Come back in a couple of weeks," you push D up two weeks in your plan. If the project manager says they are dropping F, you take it off your plan.

After I set up the new Weekly One Note page, I sent an email to the team, reminding them to update their weekly goals. This is important in the beginning, but after some time it becomes part of the standard operations. You can even give up the responsibility of making the new page, and say, "Whoever opens the notebook first on the first day of the week will add the new page."

During our weekly one-on-one meetings, we had the table to guide our talk. During my one-on-one with my manager, I was able to answer her questions from the plan. I was also able to ask specific questions.

You could do the same with an Agile system. All the tasks of a project are written on separate cards. Writers move tasks between the Future, Now, Next, and Done columns. If you choose this system, make sure each Done card has a check (no issues) or an explanation of issues faced during this task. The drawback to this system is there is no room for unplanned, outside tasks that writers typically must do.

A bigger disadvantage is if the cards are physical (not on a software), you must go where the cards are, to see the status. Instead of a table on your laptop with all the data you need, you must take notes, and you miss out when dynamic changes happen. A physical card system only works if the whole team sits together and you are close by.

The problem with using a software weekly planner system that is independent of all other tools is added overhead. I have found greater success with a weekly plan joined with the tracker. We'll get into the tracker system soon, but here are the basics of integration. The writers enter their weekly goals in the tracker. They mark the goals as "goal" in the **Duration** column, and they enter the last date of the week in the **Due date** column.

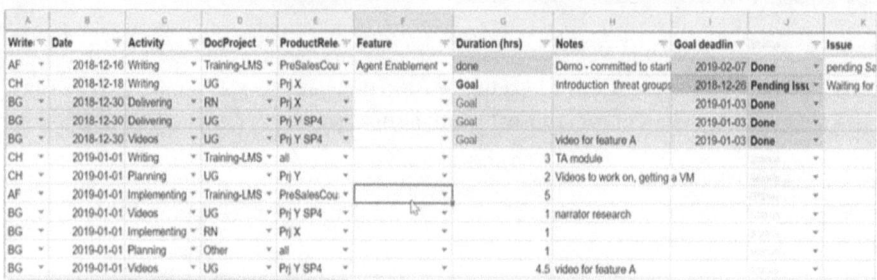

If writers know of a goal due next week, or if a goal is pushed up, they change the due date to the last date of next week or a future week. Non-goals have a number (hours, or a fraction of a day, depending on which your team chooses to use) in **Duration** and no due date. Now you can filter to see this week's goals and future goals. If a goal was completed in one day, the writer replaced "goal" with the actual duration, completing both the plan and the track in one effort. Because only goals have a due date, your filter still includes this goal.

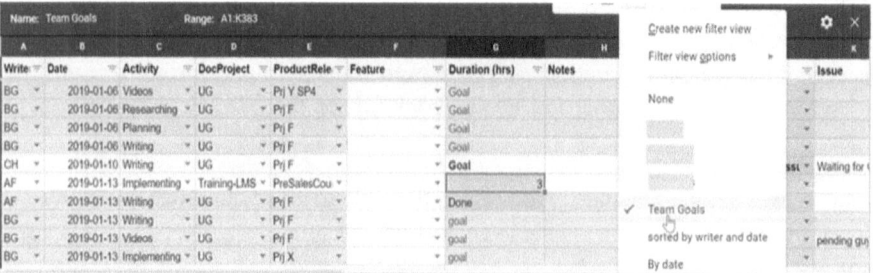

Add an Issue column, to explain briefly why a goal was not completed as first expected. The advantages of a joined goal and tracker are: plans are part of tracking and not added overhead, both planning and tracking are faster, and plans and actual work are easier to compare. The downside is your Time Tracker will be heavier. After one quarter with five writers adding daily to one spreadsheet, it might be too heavy to continue. You will need to look into a database or scripting solution to keep the data consolidated and accessible over many files.

You can keep weekly goals in a change request system, such as IBM's Rational ClearQuest, JIRA, or Bugzilla. It works if every task is recorded and broken into weekly tasks, with due dates, and resolved diligently. You can filter for this week's tasks, next week's, and overdue tasks, to see your team's weekly goals and status. This system does not work if your team does not make a record for every task. I have not yet met a writer who remembers to update every JIRA issue when it is completed or the due date changes, but if this is the system you choose to use, coach your team to use it.

With each solution, you must record the problems you faced during the work. Make sure every writer enters issues as briefly as possible. The status of past goals is less important in itself, and very important in its relation to other issues. Do not allow writers to write novel-length explanations of why a goal was not met. Do not let writers use this to record excuses. The reason for recording issues is to find solutions.

Weekly Obstacles Become Team Opportunities

At the end of a year, or quarter, go over the issues - the things that blocked your team from meeting goals - in the weekly plan.

1. Group issues by the similarity of type or cause, keeping track of how many times each item of the group was mentioned.

2. Make a list of the issue groups, with their counts. You can sort the list to see which issue types came up most often.

3. Schedule a meeting with your manager. If you can find a solution to even one problem, you have gained a lot in terms of improving your team.

Example: Handling Illnesses

In the first year, the issue that impacted the most weeks was Out Of Office. After I showed the team the groups of categories and mentioned that OOO had the most impact, some writers approached me with health issues.

When an employee comes to you in confidence with such news, listen. Do not give advice. Do not interrupt. Do not close yourself off, physically (moving away or crossing your arms) or mentally (it will show on your face). Just listen. When it is time to answer, be compassionate and objective,

"I have an issue. I have a chronic condition."

"I'm so sorry! This must be so difficult for you. How is your family handling it?"

This is important. You need to know if the physical illness is affecting the person's state of mind, and if it is causing more issues at home, which will also impact their work.

... [listen]...

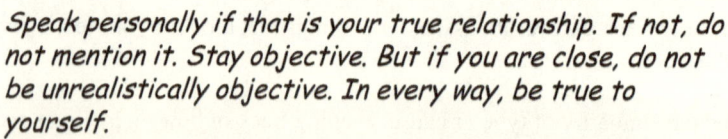

"Thank you for sharing with me. It's important that I am aware of such things, [besides being your friend and caring about you] so that we can adjust expectations accordingly."

Speak personally if that is your true relationship. If not, do not mention it. Stay objective. But if you are close, do not be unrealistically objective. In every way, be true to yourself.

"If you do not want me to share this information with anyone else, I will not. But I would like to know what the company can do for you in this time. Would you like me to speak with HR about this? Or, you can feel free to go to HR directly. Let me know if you have difficulty setting up a schedule."

It might go without saying, but it is very important: do not share personal information without permission!

I've seen writers choose to take unpaid leave for excessive absences. Together we decided they would take on less full projects and more small projects that wouldn't be impacted by frequent or long absences.

Other writers choose to work less. If HR offers, writers with chronic illnesses could opt for half-time, with accumulated insurance paying the difference (in some countries and employment conditions). When this happens, you should request a new half-time employee to cover the difference in effort.

And sometimes an employee decides health is more important than career. If this happens, work with the employee and HR to begin a well-planned, easy exit.

If you do not ask, if you do not care enough to notice, some employees will have no incentive to bring it up. How many would continue to work, to the detriment of their health?

Example: Keeping Overachievers

Absenteeism can be an indicator of boredom. If a person thinks they learned all they can learn in this company, and they moved as far up as they can go, you will soon lose them.

If we hadn't been tracking issues, I wouldn't have known that my top writers were bored and ready to look for a new challenge in a different company.

With tracking, I was able to ask writers, "Is everything OK? I'm asking because I notice that you've been out sick often lately."

That one question streamlines the problem-solution flow that keeps your employees happy and productive. It lets you show and follow through on real human care when your schedule is too intense to notice patterns of absenteeism without keeping track.

I asked two bored writers if they would like to switch areas. Two newly enthused writers came to work for the next tens of weeks totally inspired.

Example: Underestimations

Let's look at another issue: underestimated effort required to complete a goal.

First, do some investigation with your writers. How many times were the goals truly underestimated, and how many times was the real issue that they took an unplanned task that impacted their original goal? These are different issues, with different solutions, so must be tracked correctly.

For the remaining underestimates, why did this happen? "It just took longer," is often the answer. Filter for these issues. Is there a pattern?

Maybe the tasks that are always done after lunch take longer. The writer is tired after a full meal. Discuss with your writer how to better organize their time for difficult tasks.

Maybe certain tasks require resources from other teams or physical locations. If projects that require server room access, for example, always take "longer than expected", change the estimated effort. Find the average time required to get access and do the server room task. Use this as the estimate for similar tasks in the future.

Notice that **over**estimates are never an issue. No one records that they overestimated effort. They horde those extra hours to relax or to gain a head-start on other tasks. This might not be an issue for your writers, but it is an issue for you. Do not let writers add buffers to plans.

Project managers already added buffers, believe me. If a customer demands a feature by Q4, management makes plan to have it ready by Q3. The VP of R&D tells the developers it must be done by the beginning of Q3. QA says they need time to test, so the first drop must be by the end of Q2. That is the deadline you get. You are told to have the documents done by the end of June. But the customer will not get the product until December.

If you add buffers to your plan, you say you can have it done by mid-July. Now there is an issue in the plans, where no issue actually exists. The complete plan has room for unexpected problems (the F1 map does not integrate, the doc is corrupted, the network goes down for a day, people get sick or go on vacation. All of that can happen, and you will still have the docs ready before release). If you try to plan with buffers, you start your journey with an incorrectly scaled map, and everyone suffers.

 **The correct way to take into account known possible problems is to add a Risk Management Plan to the Document Project Plan.**

Your goal is accurate estimates. They let you correctly prioritize tasks and make trustworthy writer assignments. When SMEs do not or cannot deliver on time, you can show the real impact that it has on your team. If you ask for reviews in three days, and nothing bad happens if reviews are returned in six days, no one will ever return a review in three days. But if your schedule is based on accurate estimates, the impact of non-delivery is very clear.

Example: No or Late Reviewer Response

In 2013, one of our biggest issues blocking our weekly goals was "No (or Late) Response from R&D". To resolve this, my manager and I came up with a solution that the team implemented.

When a project started, we scheduled the day of when the documentation or topic will be sent for review. Then we added a task for the SME, to review the docs and return comments on a due date. The scheduled inter-team commitments are on a presentation or other plan, very dry and objective, using a template. In this way, every writer can communicate the demands as company goals.

If we do not get a commitment from the SME or managers, we escalate with, "We must delay our commitment to this project until there is a commitment from other stakeholders for documentation review."

If the review is one topic at a time, it usually gets a faster response than a review of a complete PDF. SMEs were more willing to commit to the review the next time it came up because it was easier than expected. Each feature owner reviewed the topics they knew about, not the complete chapter or book.

We also found solutions to send diff reports, to show what changed in the documents since the last time the reviewer saw it. Many SMEs could not read the complete documents. For some, English was not their native language. Others did not have a priority for the time it required. In a diff report, do not mark typo corrections and obvious language changes. You can say that many changes were made to improve the usability of the document, without marking every line.

For the final reviewer, the Product Manager, Project Manager, or Release Manager, let them know the Final Draft task is to answer: Is this a good document, in whole? Can you sign off on this document as part of your product?

Consolidate Debt

After you aggregate the issues of a period, give your team the list. Now writers can explain obstacles with "Issue 5" instead of explaining everything that is specific to this one case. This saves everyone time.

Another advantage is that you can see if some of the original issues are not repeated. From the next period's aggregation, you can whittle down the list to ten issues. For each aggregation, you decide which issues can be solved outright, which require a tool, which require a change in methodology, and which cannot be solved but can be better predicted.

The opposite of problems is successes. The weekly plan in itself offers small wins. Every goal that is met is a success. At one time, I added a column for Successes to the weekly plan, but it wasn't used. Completing the goal was success enough. Some writers (myself included) enjoyed putting a green checkmark next to each task when done. Others chose to touch this plan as little as possible. It didn't matter. In the end, each writer had some small wins from last week to celebrate on the first of every week, usually as the first thing they did. They all reviewed the year's weekly plans when they wrote their self input for performance reviews and saw how much they had completed each week.

Summary of Weekly Goals

To summarize, the weekly plan gives you:

- Teamwork, if everyone can see the plan. If a writer is out, another writer can see on the plan what is left to do and what the goal of the week is.

- Status reports at any time, with solutions to delays before they become bigger problems.

- Continuous forward movement because no delayed tasks are dropped and forgotten.

- Quick assignment of new tasks to the writer available to handle them.

- Constant improvement, when everyone records issues as they happen and you aggregate, analyze, and solve the issues.

- Small wins that lead to higher morale.

One last note about weekly plans: people own their plans.

Do not write the goals for the writers. Each person writes and updates their own goals. You can ask for updates. You can remind your team to write their goals. You can ask why next week's goals are not there. But if you take over someone's weekly plan, that person will see it as overhead, for your benefit, not his or hers. They will stop updating their plan and reporting on issues. If in a meeting with a writer, you decide together to change the plan, the best course of action is to give the keyboard to the writer. Better (I hate when others use my keyboard), if the writers use laptops, make sure they bring the laptop to the meeting, to update the plan themselves.

In the past, I suggested that writers update their plans when they return to their desks after the meeting. But a strange phenomenon often happens. They forget all the decisions, including the one to update the plan, between the time they leave the meeting room and the time they sit down at their stations. For writers without laptops, I asked the writer to send a "meeting minutes" email, immediately after the meeting, from notes they took during the meeting. I suggest you keep your own notes, in quick shorthand during the meeting, and compare the writer's minutes with your own.

7.3. Team (and Other) Meetings

The meeting is the most expensive resource you can use. It requires the time of every person in the meeting, the time for follow-up action items, and the time for overhead. For example, if you call a meeting of your team of ten for one hour, it costs you 10 working hours, plus at least 1 hour of prep time, plus at least 1 hour of follow-up time. The one meeting costs at least 12 hours of work time, or 1.5 working days.

In the 1990s, we had review meetings. QA and R&D would join tech writers in a room. Each person had a printout of the document.

On Judgment Day, the Trees will stand against the Technical Writers and we will be held accountable for every unnecessary printout.

We went through the document, covering all the comments and changes they had requested, and explained to each other what we really wanted. It settled some conflicts that were going round and round in email threads. But consider the cost of just one of those meetings. It would quickly add up to hundreds of working hours. That is wasteful. If you consider the cost, and still decide a meeting is the best solution, make it as efficient as possible.

7.3.1. Team Meetings

Have the team meeting every week, but do not set the recurrence before the first meeting. Recurring team meetings cause some people stress. And those people will be the employees you value the most. They have enough experience to know that routine meetings are too often boring, wasteful, and pointless. Of course, your meetings will be the opposite: engaging, efficient, and meaningful!

1. An hour or two before the meeting, put your thoughts in order.
   What are your goals for this meeting? What is your vision? What negatives can you anticipate, and how will you react to them? Do not put these ideas in slides, and do not try to memorize a speech! Use these notes to lead you when you lead your team in this change.
   In the beginning, you might have nothing more than a round of who is working on what. Do not waste too much time on that. Do not let one writer monopolize the time with excessive detail.

2. To make sure you are not wasting time, set an agenda for the meeting.
   I tried a few different ways to give everyone the agenda. At first, I sent the agenda earlier in the day, but the agenda often opened discussions and even arguments

through email and phone calls, when I actually wanted to use the topic for brainstorming. When I stopped the early send (MS Outlook > Delay Delivery) and sent the agenda AFTER the meeting, it worked much better. Employees who don't like to talk in the forum, or who didn't have time to say everything, were able to make their points clearly in reply to the after-meeting email. Actually, subsequent questions and discussions happened less and less because I didn't have to waste time explaining my intentions.

Agenda Topics:

• Updates and important information (communicate corporate news and policies, or let team members update everyone on changes to their projects)

• Decisions from the last meeting or internal debates to lead to a decision

• New topic or innovation brainstorm

• Review of style guide, one piece at a time

• Review of controlled language (CL), one or two examples of writing to fix

Review is an important part of maintaining a CL. If you use examples from your team's work:

• Do not put people's names

• Do not give examples that easily show incorrect writing

• Try to find examples, or ask for examples, that require brainstorming and discussion

3. Choose a format for information delivery during the meeting.
We tried different formats: no presentation (verbal only), PowerPoint presentations, printed agendas at the meeting, and document projection to a screen.

• Verbal-only is best for the short meetings, where only updates are on the agenda, and there are only a few. This is an atypical meeting, that causes an interruption in the routine. And the decisions or information are forgotten if you do not make a special effort to document them.

• Presentations took a long time to prepare, and often put people to sleep.

• Printed agendas became an asset, but are wasteful. Many writers pinned certain printouts on their boards, but each had their own (and sometimes conflicting) notes written on them. It required another meeting to synchronize everyone's notes.

• The agenda as a one-page doc, on the screen during the meeting, and in the email afterward, worked best for most meetings.

4. Delegate duties during the meeting.

Make sure everyone has a role. The delegation of duties is necessary to keep everyone engaged.

For example:

- One person is responsible for note-taking (usually the newest member of the team).

- One person is responsible for timekeeping.

- One person is responsible for updating the style guide, and one is responsible for updating the dictionary, when the meeting results in a decision to change one or the other.

Technical Writers – Team Meeting
13 March 2019

Updates and Reminders
- J. Mason promoted. New SME for Project X is Merry Stimes.
- Please remember that streaming music is not allowed during normal work hours, due to bandwidth issues

Last Meeting Decisions and Open Topics
- For our How-To articles, we will use lower-case for "to"
- We agreed we can each meet the 8-hour expectation for videos
- Alex report – should we use "therefore", "thus", or can we do without?

Review: Style Guide
List items do not have punctuation, unless one item has more than one sentence. If one item has punctuation, all items have it. Keep it consistent in multiple tables on one topic.

List items always begin with an upper-case letter.

Review: CL
Example: The remaining events may end whenever the purpose of the event is achieved, ensuring an appropriate amount of time is spent without allowing waste in the process.

Topic: New Google Sheet for Support Notes to Write as Troubleshooting Chapters

---

When the team is under stress or seems to be at a point of low morale, use this time to give them a treat (chocolate goes a long way with most) or to show corporate videos or commercials. But remember that each person on your team has tasks and deadlines. Morale-building should not be the entire meeting, and it should fit everyone's personality. If you have an employee who hates acting, don't force a team-building skit on them.

Use this time to show your appreciation of their hard work. Tell them about other teams who are having a hard time. Brainstorm about ideas to fix a problem or how to fix a relationship with a particular SME. It is counter-intuitive, but an hour for your team to lock themselves away from everyone else, is part of what brings them closer to everyone else.

They are not a group of tech writers who stand outside the corporate culture. They are a committed, integral team inside the corporation.

7.3.2. Brainstorming Efficiently

Unguided brainstorming without follow up is a colossal waste of time. With a team of seven and yourself, an hour-long meeting is one 8-hour working day. How can you ask your team to put in 100% effort of every day, if you waste a whole day? On the other hand, an hour brainstorming can result in awesomeness well worth one working day.

Jack Welch, the CEO of GE 1981 to 2001, laid out rules for strategic planning that become a template for deciding if a brainstorming session is a good idea.

• Is there a change coming that you are not prepared for?

• Is there a new trend in documentation that you could use to improve your competitive edge?

• Is the reality of your team out of sync with your vision?

If you say yes to any of these questions, you can bring in your team to find solutions.

To prepare for the brainstorming session:

1. Have a clear goal for the brainstorming session. Write it down.

2. Create an agenda: *Problem, Solutions, Action Items.*
    Solutions and Action Items are empty placeholders. But for Problem, describe it clearly in one sentence, in terms that lead discussions and bring in motivational passions.
    For example:
    "We are not able to fix PDFs in 48 hours for customer feedback."
    "Our team is going to move to a department with less budget."
    "Our competitor just got 4 stars for documentation, when we got 3, which gave them a higher score in a public product test."

3. Set a time for the session, as short as possible, with every minute targeted to the goal.

    a. Set a tenth of that time to discuss the problem (make sure everyone is on the same page).

    b. Set half of the remaining time to brainstorm.

    c. Set most of half the remaining time to rate ideas and assign action items.

    d. Leave the last 5 minutes to set up a schedule for follow up.

For example, a session of 30 minutes has one minute to start, 3 minutes for the problem, 11 minutes for brainstorming, 10 minutes for looking at the tasks of the top solutions, and 5 minutes for the end.

4. Assign roles to your team (first in your plan, before you tell them):

   - Mediator: An impartial compass who does not judge ideas, gives everyone equal speaking time, blocks negative reactions, and calms interruptions. Requires: professional attitude, willing to go with the decision of others on this issue.

   - Note-taker: Makes visual plans in a real-time document that everyone sees during the session. Records action items with task, owner, and due date. Requires: Types or draws quickly, does not get flustered with pressure.

   - Time-keeper: Has the brainstorming agenda and sets alarms to sound at the end of each part.

   - Updater: Updates the style guide, dictionary, or methodology after the meeting with decisions made during the meeting.

5. Prepare the Mediator and Time-keeper before the meeting, with the agenda, goal, and their roles.

6. Prepare the Note-taker with their role. Let them pick their tool to make a real-time, visual plan, that can easily be updated and is always available.

Be ready for issues. Stick to the agenda. If you run out of time, schedule a continuation meeting. Make sure everyone agrees with the notes, which will lead the next meeting. With time, you will be able to better predict the time your team needs.

To run the session:

1. In the time you put for this, explain the agenda in less than one minute and then explain the problem.
   If there is feedback, make sure the note taker updates the problem where everyone can see (if possible).

2. Call for possible solutions.
   Make sure you abide by the Mediator's instructions. Do not pull rank.

3. When the Time-keeper calls time, have the team rate the solutions by how they fit your customers, how it compares to competitors, and similar criteria.

4. Break down the best-rated solutions for obvious high-level action items. Have the team rate the solutions by difficulty, cost of investment, and whether they need help from outside the team.

5. Vote on a solution to implement.

If you cannot get 100% agreement, ask the Note-take to send everyone a link to the plan. Advise your team to discuss the solutions. Set a follow-up meeting to continue.

Do not give details to anyone outside of the team. You do not want to promise a change or solution before it has been tested.

6. If the team agrees on a solution, break it down to Action Items, with due dates.
    Best Practice: Assign a person to each task. Describe each task as dependent on other tasks. Set a schedule according to the dependencies. The due date is the end of the last task. On this date, you'll have a pilot or prototype. Set deadlines for testing and for communication of the solution to your manager and other stakeholders, but do not communicate the problem or solution to them yet.

7. When the Time-keeper calls time, schedule the follow-up meeting.
    This can be a status meeting, if the solution will take time. It can be a debrief, if one person implements the solution quickly. Or maybe you need the meeting to schedule time for everyone to do their action items. Schedule the implementation of this solution. Without deadlines, nothing will move.

In the next one-on-one meetings, update everyone's schedule for their action items on the new solution. Follow up. Ask about status, to make sure this on-the-side project is not dropped during daily work.

7.3.3. The 1:1 Meeting

When you have the weekly goals in place, review it before you meet with the employee in a one-on-one.

Do not skip this meeting! Do not go into it blindly. Know your employees' issues and be prepared. Sometimes they need a sounding board. Sometimes they need real solutions. You will never know if you do not meet, and these meetings will be ineffective if you do not prepare.

The one-on-one meeting with each employee should be scheduled for the same time every week. It should be as close to the start of the week as possible. The harder it is for you to get through this meeting, the more you want to cancel it, the more necessary it probably is.

Do not hold things for the meeting if they are important. Give positive feedback immediately, and give negative feedback in a meeting dedicated for that. You do not want employees to fear this time with you, so don't make it about the negatives.

If you manage remote employees in a different time zone, you might need to meet every day. That depends on the employee's needs and their complete communication skills. If they do not keep you up to date with issues and status, get it proactively.

Know your employees and the best way to reach each, based on their personality. We all have different degrees to which we are comfortable getting feedback and help. In my experience, there are "types" of writer personalities. This section gives tips for communicating with different personality types in a one-on-one.

Meeting with the Motivated Veteran

Writers with experience who are confident in their abilities and hungry for more success, often see feedback as constructive. These writers will usually be quiet in the beginning. They wait to see what you have to offer them, to help them improve or to challenge them.

If you miss the meetings too often, confident that the motivated writers are self-reliant, they will think you do not care. If you give honest and objective feedback, you build trust. They will bring issues to you that you should know about and which you can handle faster or better.

Example: Lending Your Authority

I had a writer once who was absolutely amazing, leaping from the starting gate. She learned on her own and often delivered quality deliverables ahead of time. We met every week, during our scheduled one-on-one. Sometimes we looked at the weekly plan for a minute or two, agreed, and then talked about her weekend for ten minutes. She knew I cared about her work and trusted her to do it.

She trusted me to do mine when the time came. And it did come. A project manager was giving her grief. She came to me. I ran a conflict resolution procedure that took about ten minutes with the project owner, and a week later, another ten minutes with his manager. Twenty minutes of my time, with authority that she didn't have, resolved my writer's issue.

Example: Juggling for Inspiration

During this meeting, and as often as you can, make sure your writer is happy and challenged.

The best writers get bored. People with ambitions to make a difference or to reach their potential, will reach expertise in an area and be ready to look for something new. Make sure that new challenge comes from within, on time, or your writer will find it elsewhere.

I had a number of excellent writers. The company had a plethora of product lines. After two years on the same product line, one of these writers said she would like to do something new. She was thinking about asking to join a marketing team, just to get a different style of writing on her CV. I asked her to wait a day.

In that time, I spoke with two of the other top writers. I looked at the productivity reports and spoke with the R&D managers. Then, I asked each of the three writers what they thought of my plan: juggle the product lines between them. The first was quite relieved to stay on the team with new and interesting work. The second asked to remain in his current area – he wasn't bored and enjoyed being an expert. The third expressed her fear of maintaining productivity while learning a completely new technology. I suggested that if we let fear guide us, we were going on the wrong path. This was a risk worth taking. If she didn't like the new area, we could re-arrange again, with little loss or penalty. We agreed on expectations.

The juggled writers showed a leap in motivation that lasted much longer than I expected.

Let me give an epilogue to this story. A few years later, each of those writers confessed they had been approached by competitors. Each said something similar to, "I couldn't be bothered, even for more money because I don't want to leave this team."

Meeting with the Burned Writer

Writers can get burned by their managers. Some might ask for help and be punished for asking. Others might have their objective self-input used against them in a permanent record. Some people are simply uncomfortable getting help. Some writers received harsh criticism on their writing early on, before they learned to distance their sense of worth from the quality of their work.

Your first reaction will be to let them off the hook. It takes a lot out of a manager to give constructive, consistent feedback when the recipient does not want it. Do it anyway. Do it a lot. In fact, make sure the weekly meeting happens and spend more time with that person.

Remind yourself of your goal before this meeting. Getting status updates is the lesser goal. Helping this writer improve as a writer is the main goal. Building a relationship of trust is a by-product.

1. Start by putting them at ease. Talk about the weekend or a leisure activity they are interested in. Take a real interest in this person. Get the writer talking.

2. Give your feedback. Be practical, direct, and unemotional. Focus on facts, for both positive and negative feedback.
    If you give positive feedback without facts, they will think you are manipulative. If you give negative feedback without facts, they will become defensive.

3. End with a call to action. Do not tell them what they did wrong. Ask them what they think they can do, to do better.

"We estimated that we would be able to deliver this smaller project in two weeks. Tell me why it needs more time."

"The project is developed by a contractor. Our product manager committed to sending review comments in three days, but she says that the contractor must review the docs. The contractor is not responsive."

"What can we do to fix this?"

"We can ask the product manager to handle this. Or we can deliver what we have without the contractor's review. I guess it's up to the product manager. She already has the PDF, so we can call this project closed."

"What can we do to prevent this in the future?"

"We should keep a note that for projects with contractor SMEs, we need to make sure they understand the doc review requirements. We should tell the product manager to make sure this task is in the first negotiations. And we should get to know the contractor, so that our request for review - or information - doesn't take them by surprise."

Meeting with the Worrier

There are underlying factors that create constant anxiety in writers. I have found these to be the most prevalent: ageism in hi-tech and newbies. Tech writers can be very anxious when they realize the job is more involved than they were led to believe. This profession is often advertised as "All you need is English" or "Well paying job for the over-50s".

If it is not clear yet, those are bad ads. The first is a downright lie. The second is not true for everyone.

You may get an engineer who was made redundant and thinks this profession is a good compromise. Without tech writing training, and sometimes even with it, these writers are in for a shock.

A common symptom of the Worrier is speaking out too quickly. The anxious writer will try to impress immediately. They are overly worried about their value to the company. This blocks acceptance of critical feedback. You will see these writers deliver hundreds of pages that explain what the user sees on the GUI, with very little added value. These employees never miss a meeting. They wait outside your office five minutes before it is scheduled to start. They come or call multiple times a day.

Your goal with the anxious writer in the one-on-one, is to put their fears to rest and get them focused on the tasks. Fear and anxiety produce physical reactions that are similar to true effort. But I know the difference. I tracked productivity (words written, graphics created, videos completed) under normal stress levels and again under extreme deadlines, both for myself and for writers. I discovered a clear summary: confident writers have the same productivity or less while working under stress, and anxious writers have consistently less productivity and quality. When you are dealing with fear, your energy is spent on the fear itself. You think you are doing more, but you are not. You try to take shortcuts, which usually cost you time and quality. And the adrenaline rush can become addictive. It is fine to work with a sense of urgency, if you can work methodologically.

To help your writers succeed, you must help them overcome their fears:

1. Start by putting them at ease. Plan to spend more time on this point than with other writers. Make sure you have this writer sitting next to you or perpendicular. Avoid the face-to-face, over the boss' desk encounter.

2. Hear them out – to a limit.
    Often these writers come in with a prepared speech or jump in with ideas or criticisms after a minute or two. Discreetly be aware of the time. It is very important that you not interrupt them with statements or questions that can be seen as critical. But when the time is up (you decide what the limit should be, maybe try ten minutes), stop with something similar to:
    "Sorry to stop you. I think I get the idea. Tell me, why should we do this?"
    "How is this better than what we have in place?"

"This sounds like a new methodology. It deserves a full engineering process and change management. Do you have the time for that?"

If you hear them out, they learn that you appreciate their value. If you do not let them have a voice, you will have a conflict. They will continue to try to impress you with "above and beyond" before they are simply meeting expectations. And often, if you ask "why", their search for the answer will lead them to understand the truth: they don't want a new or different task. They are simply trying (too hard!) to get a moment in the sun.

3. After their prepared speech, get the status of the tasks from the writer.

   Make sure the weekly goals are not too optimistic (under-estimating actual work required). Make sure the goals do not have buffers (including risk management in the schedule rather than in a mitigation plan). And make sure the writer knows the priority order of tasks.

4. Give the writer a list of tasks that can be done if everything else is done, or if one of the weekly goals is held up (SME sick, review not returned, network down). Idle time feeds anxiety.

5. Then, it is time for feedback on last week's goals.

   Be careful. These writers require, in the pit of their stomach, constant reassurance and validation. At first, do not wait for the meeting. Give them the constant feedback they need immediately. Make sure you answer their requests for approval of texts as quickly as possible. When you edit their writing, make it clear what is a mistake that must be fixed and what is a suggested change for style. When possible, bring feedback to the Vision and your definition of quality. If you define quality as "relevant", explain that "comprehensive" is not a part of the team vision. They do not have to document every feature. They document what is relevant to the user.

   On the side, privately keep track of the issues on which you have feedback. When you have a list, or after a month or so, find the issues that are important to you. Find one or two on which the employee consistently gets approval for the task without help from others. And find one or two on which the employee is not improving. Then prepare for the next one-on-one.

   "You are really doing well on A. You don't need to ask me in person to approve this part of the project anymore. Now let's look at X. How do you think you are doing?"

   Let this be an opener for gentle feedback on specific ways to improve an issue. This is a dialog. You guide it towards your prepared goal, but the writer must discover the solution on their own. When the writer has a solution, set a target and a deadline for the results.

6. Give a few minutes to coaching on the zen of technical writing.

   "If you find yourself jumping from task to task, be aware that you are working from fear. Stop, take a breath, and bring yourself and your work into focus."

"There are different tricks for moving on to the next thing while you are waiting for feedback. Do you want to talk about them?"

"When you get my comments on your next writing, before you open them, remind yourself, 'It is the results of work that are being judged, not me.' You are a worthy person and a valuable employee. Don't judge your SELF by your work. Distance yourself from the critiques. Learn and improve your skills. But remember no one is judging you, as a person."

7.   End on a positive note, on a You-can-do-this statement.

8.   Close the meeting.

If a "Thank you," doesn't work, let your body language tell the writer that the meeting is done. Sit up straight and put your hands on your keyboard. Without a clear and consistent sign, the anxious writer will fall back on doubts, manifested in the desire to continue the meeting. If you get in a habit of letting the meeting drag on, it is hard to break. Cut off at your time limit, and dismiss the writer gently: "We'll talk more later."

Example: Gentle Coaching

"OK, you have a solution for making sure your writing is proofread before anyone sees it. Let's give this a try. You run your idea and send me your work when it's ready, for the next five features, that you will do in the next two days. If I don't find spelling or grammatical errors, we'll call it a success. You will continue to use your solution and you won't need to go through me."

And then: "I'm here for you. My door is always open when I'm available. But I cannot always be available when you need help. It will be less frustrating for you if you can be more independent. I don't want to drop you off in the deep end, so let's go slowly."

Straight-forward feedback that works for other writers can offend the worrier. Prepare for this writer before you send feedback. They need gentle coaching.

Example: Communication Coaching

I often give feedback to interns and new writers on writing emails.

If their emails are too long, unfocused, or miss the point of the thread, I will tell them: "Keep it short. Keep it simple. Stick to the one point. If you have more to say, do it in another thread. If you must, ask for a meeting."

Out of maybe ten times I have done this, three who showed constant anxiety symptoms started to CC me on every email, without changing the way they communicate. I kept up the feedback until they showed consistent improvement: short, focused emails. Then I said, "I really like what you have been doing. You don't have to CC me anymore, unless you want me to step in."

7.3.4. Representing Your Team in Meetings with Your Manager

I suggest that you put aside an hour to prepare for this meeting.

Review the weekly plan. Do not present it as-is. Summarize the status, issues to report on, and questions you have. Have the whole plan available to you, on your laptop or printed out, to be able to answer questions quickly and accurately. Review the current project plans. Make sure with your team that the plans are updated.

If you are asked a question you cannot answer, the worst thing to say is, "I don't know." No, retract. The worst thing you can say is something inaccurate or untrue. It is better to say you do not know. But it is best to say, "I will have to get back to you on that."

Make a note. When the meeting is done, find that answer quickly and send it or bring it personally.

7.3.5. The Most Important Meeting Tip - Template for Minutes

The time you schedule to meet with your writers is time taken for that purpose alone. Do not look at your screen. Do not look at your cell phone. Maintain constant eye contact with the writer. Expect the writer to take notes and to make the changes to their weekly goals.

If the meeting results in an action item for you, follow through. If you cannot set yourself a reminder, ask the writer to send you an email to remind you (such as the meeting minutes). If the weekly is with a writer in another location, try to use video tools. Make sure chat is also available, to use if the audio lags. Ask for minutes. This is required for distant writers, and it is a good idea for all writers.

	Owner	Notes
Purpose of Meeting:		
Decisions:	1.	1.
	2.	2.
	3.	3.
Action Item:		Due:
		Priority:
		Next Step:
Action Item:		Due:
		Priority:
		Next Step:

7.4. From Vision to Expectations

You can be a team manager without a vision, but a clear, well-defined, real vision is necessary for leaders. If you know how to solve the issues of the day and week, you are a manager. If you know how to solve the issues of the day in a manner that brings an improved future, you are a leader.

Creating a vision means that you have the first ingredient for leadership: know where you are going.

To create a vision:

1. Define your purpose in one sentence.

   Why does your team exist? How does that connect with the company's purpose? Write the purpose as though it were already true. Examples:

   Our documents are the easiest to use in the sector.

   Our users are the best trained in the industry.

   We are the experts: our docs are the go-to for all product_type users.

   Our deliverables win awards every year.

   Your team's purpose is the foundation of continuous motivation. This is your foundation for morale. "Things might get difficult, but we have a mission. We might be focused on small tasks, but everything we do in a day leads to our bigger-picture mission."

   Make sure you have complete team engagement. This first step, in fact, this whole process, can be a joint effort. If you work as a team to create your vision, make sure that you have the final say and that you use your veto power. Do not drag out this process too long.

2. Expand your mission to a detailed vision of your team's optimal work and deliverables.

   I usually start with the basics of technical documentation: controlled, relevant, and accurate. Then, define each of those adjectives in terms of how you work, what you deliver, requirements, and risks. Make sure you are leading off from your mission!

3. Test your vision.

   Can you make SMART goals (specific, measurable, achievable by the team you envision, relevant to the company, timed) for one project that lead from today's status to an end result that fits your mission? If you run this test for a random project, you will quickly see if your mission is relevant to your company, achievable in a realistic sense, and general enough to be a guide for years to come. I suggest that you run the first test before you write all the tasks as goals. If it doesn't work, you can tweak your statement without too much loss.

4. Expand your vision to be a complete strategy.
   Make SMART goals for all the projects you have now and for the known projects as far in the future as you can.

5. Define granular requirements and risks.
   What do you need to purchase and how much will it cost? How much headcount do you need and how long will it take to get it? Create a risk management plan.

6. Present the vision with the goals as measurable expectations.
   I often find that a graphical document is easier to keep in sight, but a textual document or presentation summary is important for communicating the vision to my managers.

7.4.1. Sample Vision

VISION STATEMENT:	BE THE GO-TO REFERENCE FOR DATA SECURITY PRODUCTS				
	CHARACTERISTICS:	Accurate	Easy to Use	Relevant	Easy to Find
	CHARACTER DEFINITIONS:	Tested	Concise	Updated	Expected Deliverables
			Correct Writing	User Requested	SEO on public KB
				Audience Specific	Repository for Sales
	ACTION TARGETS	100% content confirmed to change for new release	80% content uses Controlled Language	100% docs filtered to show only user relevant info	100% online docs open to public have SEO features
		90% new content tested with hands-on	90% content uses Style Guide	100% user requests tested and written in 48 hours	75% Deliverable innovations driven by users or Marcom
		10% confirmed tested by QA	100% proofread and peer reviewed	75% Sales / Marcom requests tested and written in 2 weeks	100% deliverable changes for security and Marcom image
SUCCESS:	Analytics show non-customer site visits				
	Support feedback 4+				
	Positive user feedback each quarter of new release				

Enter your vision statement. Define your vision as characteristics of the documentation. For each characteristic, enter action targets as SMART goals (Specific, Measurable, Achievable, Relevant, Timed). For the complete picture of your vision, enter success measurements. Given the success measurements of the goals, you can identify specific tasks for the foundation of your team's methodologies.

The detailed vision is a graphic. It does not change often, unless you decide the vision is holding you back. The tasks, on the other hand, can change as needed. The tasks to reach the vision goals must fit the team and the reality of your resources.

The tasks often rely on primary foundation skills. If your team requires training (for example, to create quality graphics and videos) or review (for example, to integrate the style guide and Controlled Language), schedule the training and review before you assign the tasks. Schedule daily, weekly, and project time to reach success in the tasks. For example, for the task of staying up to date, you can schedule 30 minutes every day for your team to read forums and mailing lists. Then, set aside half a day a week to investigate discovered questions and to deliver tested and proofread answers.

Given the sample vision above, the table shows tasks to reach the vision. You can use these tasks as expectations.

	Task	Owner	Success
Accurate	Meet with R&D, Sales, Marketing.	Mgr	Have commitment for resource sharing and outline approval.
	Meet with QA leaders.	Mgr	Have commitment for testing docs,
	Use the products while writing.	Team	QA rejects due to unknowns only. Hands-on > 80% of topics, 100% hands-on for CLI.
	Send components to QA for review & use their comments quickly.	Team	Zero neg feedback on accuracy.
Relevant	Stay up to date with user forums, Support, competitors. Daily, 30 min.	Specific writers	Create issue for new/change, 1 every Q.
	Automate tasks with macros, XML, scripting. Weekly, 4 hrs.	Specific writers	Constant improvement of time to delivery of feature-rich doc
	Update deliverables for Usability team & surveys. Major Release, 5 days.	Team	Delivery format is easy to access & relevant (expected) for users.

	Task	Owner	Success
	Collect & implement user feedback in timely manner.	Team	Feedback investigated, written, proofread, tested, delivered in 2 – 5 working days.
Easy to use	Create quality graphics and videos on time. Weekly, 8 hrs. or less for 5 min video or 2 graphics.	Team	Get positive feedback from users and partners.
	Use the company Controlled Language and Style Guide. Every topic, 1 hr.	Team	Concise, simple writing, <20K to translate, consistent formatting.
	Proofread and Peer Review. Weekly, 2 hrs to review others' work + 1 hr to fix own.	Team	Zero rejects for source language errors.

7.4.2. No One Left Behind

Make sure your team has a stakeholder's commitment to the vision.

You might want to work it out with them all. Brainstorm, decide, and develop your vision together. You have all the advantages when the team is an engaged unit. But if there is even one person who is there to collect a paycheck, who can't be moved, who has no desire to reach personal fulfillment through work, those advantages can quickly be reversed.

To find a solution, I like to look at the issue as my high school Attitude Awareness teacher taught us. The answer to a situational challenge is one of these: change the situation, change the other person, change yourself. Before you go charging in, make sure you have a clear understanding of the situation, the other person, and what you are willing to change in yourself. You can talk to this person before bringing the idea to the team. Get the real response. Maybe that person will surprise you and be on board. Maybe they already have a great idea. Maybe that person is just uncomfortable with the forum environment or thought you didn't appreciate him or her.

If that person really is unwilling:

- Consider how to change the situation.

  Can you offer that person an out, to not attend? That would change the situation, but it is not a real solution. The one person's stubborn, static, minimal effort standards will be a spot of rot on one apple in the barrel. It will eventually spread.

  Or you can change the situation completely. Do not bring it up in a forum. Make the vision yourself and hand it down. This might be best for large teams. But you will probably have to delay engagement and its advantages until you have a win that proves your decision leads to greater things and is more than overhead.

- Consider a change in yourself.

  Does that person have feedback for you, to change yourself, your communication styles, or your attitude? Listen carefully. When you have time to look at the feedback objectively, decide if you accept it and change. Will a change in the vision and how you function solve the issue? Will your change make the vision acceptable to the team?

- Consider encouraging a change in the behavior of the other person.

  You can let this person know they have influence and power. Sometimes, it is the person's lack of self-worth that creates a negative cloud. You can give the person a chance to explain why they are not interested in this one endeavor (creating a team vision). With that knowledge, you might be able to persuade your employee to give it a try. Or perhaps a change in attitude is necessary. Discuss how this affects results and set achievable expectations for change.

"I wanted to talk to you about something important. It seems to me that you don't realize how much you influence your teammates. Did you notice that when you got upset last week about the change in priorities, it affected everyone on the team?"

"What do you mean? You can't blame me for that!"

"I'm not blaming. I think it's cool that the rest of the team appreciates your abilities. Because you are such an important part of the team, and because you have such a powerful personality, your words and moods influence the attitude of the team. Now, the question is, how are you going to use that influence?"

"I'm not doing anything on purpose. I don't want to manipulate anyone. I just want to do my job."

"I'm saying you influence everyone, whether you want to, or not. You do your job, but how you do it, makes a difference to all of us. For example, in tomorrow's team meeting, we are going to look at the team's vision. If you are negative, everyone will leave the meeting with a bad feeling that will influence the rest of their day and their view of the vision in the future. And it could even ruin our chance to get improvements, like more headcount or that new tool we were talking about. I would like to see you being a leader in our vision. What can you do to make that happen?"

"I don't want to be a leader. I don't have time for anything more than what I am already doing."

"This won't take up time. Being a leader in this case means that you help decide the direction of our team vision. For example, when we talk about the things in our workflows that we want to change, you can simply be honest with your opinions. Talk about what can be improved. And when we talk about an overall vision, bring in some of that power. What is important to you, that we should include?"

"Well, it's really important to me that you stop asking us to work overtime."

"Great! Let's make this an action item. We will update our vision to say that we deliver on time, without overtime. How would you rephrase your requirement to be positive?"

At this point, you should see a change. They realize that you appreciate them, that they are not powerless. You might have to be silent now, while they work through a personal revelation or talk about things in the past that made them think badly of conceptual changes.

Some people get really excited. Finally, their chance to make a difference is here! You might have to calm them down a little. Do not let them go to the vision-definition meeting thinking they control everything. That will just disappoint them and push them further into the shadows of mediocrity.

Others get angry. They really do believe they are doing enough and are insulted that you push for more. If you can, let that person talk it out. Really listen, and find out where this is coming from. Ask questions without interrupting. Ask to gain a better understanding, not to point out a fallacy. Then, ask if they have ideas to improve on the situation or if there is anything they want from you. Maybe this is simply the situation you have for as long as this person is your employee.

I think one of the worst things you can do is ask that this person not join the team discussion. That can only end in insult, and your team is going to hear of it, probably even before the meeting starts. This will encourage an environment of fear and punishment, that will ruin creativity, innovation, and have an extremely negative impact on the mission statement as a tool. Instead, make a brief statement that summarizes the good things that you believe will come out of this and the hope this person can join in on the effort.

It is often surprising how well this can work. After frustrating one-on-one meetings, the negative employee who finally got a say will come to the next endeavor with powerful positive energy.

If you get to a point where you are out of ideas, ask yourself if this person can still make a contribution to the team, working as they always work. Do you have tasks for a pack-mule? Are those tasks important enough that you are willing to put this person, a complete resource, on these tasks? Sometimes, if you leave an employee alone to plug on, they come to accept changes gradually. But do check the work. If the results do not meet expectations, you will have to point out where the specific changes are to be made.

7.5. Delegating

If you are a general business manager without training or experience in tech writing, this chapter isn't for you. You will delegate all the documentation tasks to the writers.

No joke.

If you do not know how to use a semicolon, or better, why you should not, you must delegate everything. If you try to take over simple writing tasks, you will end up wasting your team's time, in the best case. In the worst case, you will burn SMEs. There is such a thing as a professional writer. If that is not you, do not go there.

Now for the hard part: tech writers who climbed through the ranks to become team leaders or group/department managers.

Do you like being a manager? Then delegate!! You are not a writing resource. You cannot fit projects in your team schedule with yourself as a backup hole-plugging fire-fighter. Yeah, maybe you can do the job faster or better — that one task. This is what typically happens to doc managers who cannot delegate: they work longer hours, they drop management tasks in favor of writing tasks, eventually (if they are not demoted first) they lose their patience for their team and SMEs and then burn out.

If you must fill in for writing tasks, make sure you track the time you spend as a writer. Know this, dear grasshopper: The more time you spend on tasks that should have been delegated, the further away will be the light at the end of the tunnel. (Just to be clear: the light is the stage where everyone has achievable tasks that fulfill all doc requirements of the company.) If you must be a writer, make sure your team vision is clear to yourself and to your managers. Bring that light closer!

Delegate! Delegate! Delegate! It cannot be stressed enough. Be the best manager you can be.

When you delegate, give the whole thing. Let your employees own their projects.

7.5.1. 1 2 3 Go!

Make sure the writer is ready to take ownership. Make the task complex enough to require ownership, but start slowly.

1. The first time a new writer takes a complete project, walk through the expected procedures.

2. The next time, let the writer do it on their own, while you verify.
    This is a great opportunity for morale-boosting small wins. "You did this plan perfectly," for a plan that you can create in your sleep might seem overkill to you, but it is just what the new writer needs to hear.

3. After a successful project with a reasonable number of changes on your part, trust the writer to own the projects themselves.

   Your meetings will not be about the next step or whether the last step was appropriate. They will be status meetings, where you learn what you require to update the high-level schedule or to answer upper management questions.

   What is a "reasonable number of changes"? Zero is the benchmark, but if that happens, it can indicate that you held on too long. That writer should have been owning their projects much earlier.

It is the same principle as parenting. It takes time and patience to teach a child to do something for themselves, but the end is worth it. You have time to do the tasks you want to do to improve yourself and your results. The writers improve themselves with new skills. And (if you do this properly, with patience) the writers gain confidence that leads to further success.

7.5.2. Engrave Ownership in Gold

When you assign a project to a writer, call them "owner" in all emails and plans. Give the owner autonomy to make decisions. Let the owner show initiative and imagination.

Do not change the project plan that an owner makes. Let the owner update the weekly plan and the project plan themselves. The only plan you change is the high-level team schedule. The owner decides when and how to do the tasks of the project, in the range of the start and due dates. It is my experience that if you touch someone else's plan, they lose the feeling of accountability for it.

That does not mean that you do not help the writer. You must be aware of the plan and the writer's status. If the writer is not handling issues by priority, say something.

7.5.3. Get off the High Horse

Novice owners need more coaching. Many technical writers with more than ten years' experience have zero experience in ownership. Prepare for meetings with the owner. Think of the behavior you want to encourage: competency, influence, intelligence.

Make sure the owner understands the importance of the project and that you are grateful you can trust the writer to own it successfully.

Planning, updating, and tracking can take time. Add *work the plan* or *work by priority* to the performance goals of the writers. Encourage plan ownership with objective performance feedback.

Be a coach, not a dictator.

"Should I call it *Glossary* or *Terms?*"

"What do you think is best?"

"I don't have a preference. I understand the decision will affect all our future deliveries, but I don't know what questions to ask."

"Here is a good question: if users have to search for *glossary*, can our search engine recognize it through spelling mistakes? I mean, if a user enters *glossery*, will the results show: *Did you mean glossary?*"

"Our Help Center search engine does not do fuzzy matches. So, we should use *Terms*. But maybe it will be confused with *legal terms*? We could add common misspellings as labels."

"That is another good question and a good solution. When you decide, update the Style Guide and let the rest of the team know."

7.5.4. Working with SMEs

Sometimes the hardest part of delegation is letting the writer go alone to meet with SMEs. If you know a writer likes to talk too much, has a learning disability, falls asleep when bored, or shows some other sign that I haven't seen yet, the writer will burn internal contacts.

Many writers do well with their contacts. Many more do less well than they think. Some think that their preference to work with humans over computers makes them a people person. Some never find out how much they actually annoy SMEs. You can coach your writers to improve their professional interactions with their SMEs.

Approach the correct SME	Before you begin a project, make sure you know the players.
Approach the SME correctly	Call first, do not barge in. Discuss what communication methods work best between the two of you. If the SME is speaking with others, do not stand nearby waiting. Ask for a meeting time.
Establish rapport	Set the groundwork for cooperative work. Act as the expert providing a service to help them succeed in their project.
Introduce their role	Explain the role of the SME, product manager, and QA. Make sure you agree on the audience and scope of the writing project. Get a commitment for reviews in a date range.
Ask the right questions	Before you ask an SME a question, read released guides and internal documents. On the first project of an unfamiliar product, find a tech writer who knows the product and run your SME questions by the writer first. SMEs do not have time to teach you the basics.
Maintain professionalism	Maintain a professional, pleasant, patient tone of voice and body language. If the SME digresses into something not to be documented, do not cut them short. Use body language to let them know: sit back, stop writing or typing, and return them to the relevant point with something like, "Can you tell me more about that thing you mentioned before?"
Make and share the plan	Make internal delivery deadlines and stick to them. You build credibility by delivering when you promise to. If the schedule shows you will not have time for everything, communicate impact. For example: "We will not be able to document the CLI until after GA. We will deliver an update to the guides after 1 day for every 2 sets of commands. If you want, we can deliver updates in stages. Let

> me know if some commands are more important than others."
>
> Send the link of your project plan to the SMEs and to the product manager. Follow the doc project workflow to lead the doc project.

Do your part	Implement comments in a timely manner and share the results. Let your SMEs know that you are using their input and not wasting their time. If you cannot use a comment, explain, BRIEFLY, why not. Do not send an email that explains everything you did with their comments. Send only issues.

7.5.5. Unplanned High Priority Tasks

Projects come up all the time. Do not dump them on your writers.

> Do you notice the recurring theme? Assign tasks with thought and schedule.

If an unplanned, higher priority issue comes up, handle it like any request, no matter who asks for it.

1. Scope the effort of the new project.

2. Communicate the impact of the scope.
    If it will take more than half a day, communicate the impact to the requester (or to your manager, if the requester is too high up to chat with).
    If the impact is acceptable, get that in writing.

3. Change schedules according to the new priorities. Do not delete tasks from your schedule. Move them up.

4. Communicate the change in-person to your writers. Make the new schedule clear to your writers. Help your writers make it clear to their SMEs.

5. Follow-through: make sure the project is done on time and that your team returns to the regular schedule.

If your team is requested to make a quick change that will take less than 4 hours, just do it. You can communicate the impact to others if you want or if there is a deadline in question.

They will not be able to do anything about it, so the discussion might not be worth your time.

"Robert, Charlie is on your project, to complete in 5 days. But we just got a request from the VP to deliver a change to project X. I need to take a few hours of Charlie's time."

"So, you are putting my project on hold? Will it be late? I'm sure the VP thinks my new project is more important than that change."

"Well, Bob, I estimate the effort will be 3 or 4 hours, so I would like to get this done. Charlie will complete your project in 5 days, but it will be by the end of the day, instead of the morning. If it takes longer, I will call you again and we can discuss."

"Oh! I'm fine with this. That's not a real impact. Just let me know if you can't deliver on the date we agreed on.

"Charlie! Come on over. I've got something for you."

7.6. Productivity

The premise of this book is that the prime objective of a manager is to help employees improve.

 These are the words of Annika Rubinstein, when she was my manager. In one sentence, Annika summed up what I had been feeling, but incorrectly defining. I had been letting a friend's advice lead me: "A manager's job is to protect her employees." But if you protect them too much, you take from them the opportunity to improve. You lose opportunities to see increased energy, morale, and productivity.

To improve, you must know where you are and where you are going. To know where you are, requires tracking. To know where you are going, requires defined expectations. To close the gap, you encourage productivity.

7.6.1. The Last Word on Counting Words

At this point, the experienced manager is holding his pen over a KPI [10] template and screaming, "Tell me already how many words they should be writing! That's the reason I bought this book!"

This is the short and incorrect answer: For GUI-based end-user applications that do not require intensive hands-on or advanced research, a professional technical writer should be able to write 500 words a day. For complex technologies, a professional technical writer should be able to write 300 words a day. That is the average of a project, when you include screenshots, proofreading, implementing review comments, and delivering.

The answer is incorrect because the question is irrelevant. Word count is the lazy manager's indicator. Counting words is a worse indicator of productivity for writers than counting bugs for QA Engineers. For QA, if you measure only numbers without indicators of complexity, the less ethical employees will fill the bug system with numerous issues that should have been combined through one solution, with duplicates (because why waste time making things easier for other teams when that will penalize performance?) and with more time on simple bugs than on bugs that could make a difference.

The situation is more complicated with technical writers. You want writers to work through research, to create virtual environments, to study similar tech from other documentation, to help with the GUI review, and so much more. If you have writers who are only writing, you still want them to write with control. You want them to consider translation costs. You want

---

[10]One of my Beta Editors suggested that I write out this acronym, but if you are not familiar with Key Performance Indicators, don't worry about it. Yet.

them to re-use chunks to ensure consistency and reduction of conflicting information. Counting words ruins all the encouragement for quality and improvements you could have brought to the table.

And a final reason to not count words: If your team creates innovative deliverables that are not strictly text-based, counting words will be rather difficult. If your writers are using an on-premises CCMS, you could get a developer to create a script that runs over the difference of words added to the back-end database. I tried this for one team. The results were different depending on the day when I ran the script, even though a time range was input. For many reasons, I found the results unreliable. It helped show trends of productivity, but not as much as other measurements.

The last word on counting words is: *Don't.*

7.6.2. Creating Expectations

The best practice is to start productivity with expectations (based on your vision, of course).

What do you expect of your writers, in terms of daily tasks that lead to the team vision? You do need measurements. If you do not count words, you need something else. For example, if the vision includes accuracy, do you expect your writers to use hands-on? If so, how much of their time should they spend on this? Do you want a minimum percent of hours per month, a min-max range, a different rage for each person?

I suggest that you have at least three categories of expectations for results, and no more than five. The categories should be about results, not efforts, and the expectations themselves should be written in objective terms of measurements. You can start with the same expectations for everyone. Change expectations as you go, according to the abilities and focus of each person.

For example, you start with a category called *Accuracy.*

Expectations - Accuracy:

- You spend at least 10% of your time testing your writing with hands-on.

- You send 100% of your topics (one procedure at a time) to QA engineers for review.

- You send 100% of your completed drafts to the project owner.

Let's look at the team with this example.

After a few months, you see that Alex is a great writer who hates going to the server room. He sees it as an unnecessary interruption in his work. Betty is a tech geek who loves fiddling with the servers, but she gets caught up in technical details and often takes much longer than Alex to complete tasks.

Alex's priority project is running fast to get a PoC (Proof of Concept) bid. The QA and project owner of this release do not have time for the docs right now. Betty's priority project is going more slowly. The project owner of that release continuously asks to review the docs in progress. Both projects are going GA in six months.

After one month of work, you see the trends. Alex is delivering fast docs to Sales that are not reviewed. Betty is going to fall behind in her schedule, and if you do not handle that, you will be responsible for a red flag on a steady project. You meet with each of them, separately.

Alex is tracking 10% hands-on. But without QA and R&D review, the accuracy of the docs depends solely on Alex, for the PoC at least. You tell him that you are changing his expectations for the next month only, and after that, you'll probably go back to the default numbers. For the next month, Alex is expected to spend 50% of his time with hands-on, and only 50% of his topics must go through QA. With Alex next to you, you call the project owner and explain that Alex must stop giving Sales unreviewed docs. The project owner must read and sign-off on the docs before they are released.

Now for Betty. Betty is tracking 40% hands-on. She refuses to move on to a new topic until QA answers her technical questions and every word is absolutely correct. You tell Betty that her expectations are now changed. From now on, she must not spend more than 20% of her time doing hands-on in the server room. She must implement comments from QA the day after she gets the comments (or earlier), and she must not send a topic for more than one review. You suggest that she put further technical questions in metadata comments (visible only to the writers) and move on. You remind Betty of the schedule and make sure she understands how much time she has left to complete the specified number of features and deliverables.

If the team did not know and have access to the written, specific expectations, and if you did not explain the change as a change to those expectations, you created an environment of fear. If you change expectations without obvious reason, your people will feel you are floundering. They will wonder if upper management is dictating the changes, and they will lose confidence in your leadership. They will wonder when the next change is coming, and if it will override this one. And thus, you lose their urgency and motivation to meet expectations.

Clear expectations let your people know the boundaries in which they can safely move. When you change the expectations, if you change them clearly and for a specific and logical reason, you continue the feeling of safety. They are empowered when they realize their own actions and results drive your changes. If Betty really wants to get into the nitty-gritty of the server OS, she is going to push up her writing productivity to be able to make the most of her 20% max time in the server room.

If you create multiple categories, each with multiple measured goals, you allow each person to shine in at least one area, and to improve in at least one area.

Make sure you show appreciation for successful expectation fulfillment, in a way that suits the person. Some people love getting attention in a group setting. Others hate it. Some people love getting gift certificates so much that they frame them instead of using them. Others hate them so much they see it as an insult. Know your people and show appreciation appropriately for each one, in a manner that is fair.

If your organization has performance reviews, use that time to update each person's expectations. Enter real and specific projects, if you know what each person will be doing. If you find yourself copying the last terms' expectations, make sure your writer is not getting bored. Sometimes, you must juggle projects to give veterans the chance to find a new challenge. Even if your people prefer to continue writing about the technologies they are most familiar with, your expectation measurements improve. If you cannot raise at least one expectation for each writer, there is something wrong. Are you asking enough or too much of your writers?

7.6.3. Communicating Productivity

You have your expectations for your employees. They are tracking their work, or you have an automated system, or both. You can see how the employee is doing. Do not wait for a performance review to let the employee know how you feel about their progress. If a person is doing well, show appreciation. This motivates. If the person is doing extremely well, show proportional appreciation. This motivates proportionally.

And if an employee is doing badly, they have a right to know. Communication of low performance is an opportunity. You might find out very quickly that a new employee is trying so hard to look good, that she is more focused on low-priority visible tasks than on trackable tasks. One word from you might set her straight. Or a writer might confide in a personal issue or illness that is affecting work temporarily. You might uncover a plot by an unscrupulous employee to gain points by stealing the work of a colleague.

 I was once told of a place where writers worked on Support cards. The software they used listed the last person to edit a card as the writer. One writer spent part of every day making slight changes to tens of cards that others wrote, and he copied cards of others to create "new" cards. The manager thought he had one amazingly productive writer and the others seemed to get worse and worse. When he finally

approached the low performers, the plot unfolded. If the approach and defense had been earlier, the knowledge base would not have been the mess it was, and hundreds of man-hours would not have been wasted.

When you communicate productivity or performance:

- Collect objective numbers on a schedule, every week or month.

- Send an email report with a template, to maintain objectivity.

- On the report, do not explain or lead explanations with questions. Keep it dry and short.

- Do not include the team average on the report. Do not include the highest or lowest numbers. Each person's report is individual.

- On the report, include the expectations. You cannot make it too clear. Do not worry about repeating yourself.

- Be prepared. Writers with low reports will often come to you, to explain. If they come to whine or question your methods, prepare to answer them calmly and objectively.

- Do not let this discussion happen on the email thread or over the phone. Discuss low productivity in a face-to-face, individually and verbally.

- General responses when the explanation is about working on multiple projects: "That is fine, but I think you can do that, AND reach the expected goal." or "I understand that other project is important to you, but as a team, we must work on priority issues. Do you want to schedule time to work on the other thing?"

7.6.4. Productivity Template

Here is an example of a productivity report for the number of videos created each month:

July	Expected	Your Work
Minutes of video / Total videos:	25 / 5	9 / 2
Days in office	20	15

The table is a template. You fill in the numbers, for each person, each time, based on objective calculations. In this example, you see immediately what the issue is. The writer is meeting expected work, when in the office. If this is a one-time issue (for example, this writer was on vacation) there is nothing to worry about or discuss.

The mistake a n00b manager will make, is to put in the email about how it does not matter this month or some such thing. Just fill in the numbers and let the writer say something, if they want.

Another mistake is to include only productivity numbers, without calculating *days in office*. Reporting days of work makes it clear that productive attendance is appreciated. And it makes it clear to yourself if an employee is missing days every month or week. What is going on there? Low attendance can be an indicator of low morale or of a discrete job search.

If you agreed to let an employee take a day off every week for study or a side project, change the expectations or discuss the feasibility of the arrangement. Did that person promise to keep up their work, or did you promise to reflect the agreement in lowered expectations?

7.6.5. Challenge or Stress?

Our careers should be full of challenges - tasks and goals that improve our skills and build our characters. As a manager, you set those challenges. And as a manager, you must protect your employees from demands that are not challenges.

If your writers have challenges, they are engaged, motivated to succeed for you and for themselves, feel their own personal growth over time, and feel they are constantly learning. Their work is fulfilling.

If your writers have stressful demands, they are anxious, worried, fearful, and frustrated. They feel they are working hard, though much of their energy is drained in the negative emotions rather than the work itself.

To promote challenges and reduce demands:

- Know and follow up on weekly goals.
  If someone does not have enough to do, give them a new project. Make sure this is not a time-filler. It must be a real project with a schedule and end result that the user can hold up and say, "This is mine!" You've heard this: "If you want something done, ask a busy person." [11] Motivation drops when the to-do list is done before the end of the day.

- Share the load of tedious tasks.
  Get to know your people and what they like to do. Some get in the zone when making graphics. Others find it tedious. If everyone on your team hates it, divvy up the graphic tasks.

- Give your people more responsibility and authority.
  When we feel that we influence decisions, we feel empowered. This leads to fulfillment, motivation, and ownership. But again, know your people. If someone is stressing over a deadline, it might not be the time to tell them to take on a new time-consuming responsibility. Then again, it might be the perfect time. Ask! Maybe just knowing that

---

[11]Lucille Ball.

you would consider making them lead in a project is enough to lift their chins up, even if they have to refuse.

- Remind your people, in team meetings and in one-on-one meetings, of your team vision, of the company goals, of the purpose of the product.
Knowing that their daily work has real importance makes the small frustration worthwhile. When we have a deadline for something in which we see no purpose, our energy and motivation fall.

- Cut out the overhead.
What steps in procedures are unnecessary? What tasks can be permanently dropped? When our work is full of red tape, we all see red.

I'm pretty sure this is my own clever witticism. But if you have heard this before, let me know.

- Make sure your expectations, overall and for weekly goals, are clear and understood by all.
Ambiguity causes misunderstandings, which generates futile efforts. Stand by your expectations and instructions. When you contradict yourself, you confuse your people. If it happens often, they will learn to not trust you.

7.6.6. Handling Negative Feedback on Productivity Encouragement

Some employees will deny their low productivity even in the face of objective numbers.

"Your expectations are unrealistic. I am working as hard as I can."

In this case, it is important that you prepare with the facts. If the person comes to you without a schedule or calls you on the phone, set up a meeting to discuss this the next day. Do not answer from your gut.

Before you meet, prepare the chart that shows this person in comparison to the rest of the team, without names. Seeing their line far beneath the team average, especially next to the chart of the average team member (without names), will do wonders. Your expectations are realistic, according to the work of others.

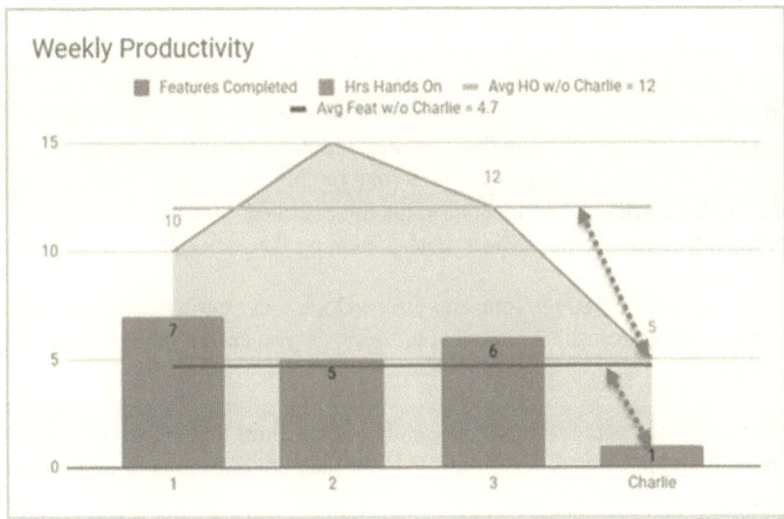

If the employee believes their work is being stolen, let them go back to their desk and do some research. This would happen if the writer knows they did X, but your numbers show X minus Y.

If the employee realizes they are performing lower than everyone else, listen to what they have to say, for no more than 30 minutes. Then: "I have to stop you. Your reasoning has validity. But I am asking you to meet your expectations. What can we do to make that happen?"

If your employee gets angry or emotional, instead of "What can we do…," you can suggest that they think it over and come back in a day or two. Let the denier accept reality and come back with his or her own solutions.

Sometimes, a writer just needs more focus on the task at hand. Or to drop lower-priority issues. Or find better methods to complete certain tasks. For example, a writer can spend a full work day on paginating and table formatting. You can make macros to make this work faster, if you are still delivering PDFs.

7.6.7. Curiosity Killed the Cat

Time and resources for curiosity during work hours is awesome but rarely experienced. Sometimes you must balance quality against productivity. There is a difference between thorough hands-on and wasting time simply because you wonder, "What does this button do?"[12] You will have to decide when curiosity can lead to better quality or innovation, and when to advise your employee to indulge during off-hours.

Good writers can create quality documents for a technology of which they knew nothing. It is lack of time, not a lack of curiosity, that drives such efforts.

I've seen writers document how to use technology in a cloud, without being able to define the cloud. I've seen writers spend weeks on documenting regular expression syntax, outputting a chapter that is relevant to high schoolers (maybe) but not to their target audience. Make sure research-based documentation fits the assumed knowledge of the users of the technology. It cannot be the writers' notes, teaching themselves.

You can cut reams of text gained through time and effort. Do not let your writers complain or make excuses. You do this because you know your users. (If you cut for other reasons, be prepared to apologize.)

Hint: look for "this means that" and "so that". These are big red flags that the writer tried to explain things to herself and lost the user's perspective. Or they are indicators of repetitive writing, also to be edited.

---

[122] Dee Dee in *Dexter's Laboratory*, created by Genndy Tartakovsky. I don't know if he wrote that part of the script or if it belongs to a different writer. I guess you could ask him.

7.7. Innovation

A nation that cannot adapt, eventually falls.

—Bobrow, Edwin E. and Shafer, Dennis W, *Pioneering New Products*
Dow Jones-Irwin, 1987), pp 1-9 (Or any credible history book)

A team that cannot improve, eventually fails.

A technical writer who cannot innovate, eventually collects unemployment.

 There are plenty of examples of companies that laid off their local technical writers to outsource. Anyone can create a manual, to varying degrees of quality, but if you can offer more, and often, budget controllers will be less willing to pull the plug on in-progress projects.

7.7.1. Change or Die

During the MEGAcomm conference in Jerusalem, 2016, I was on a panel with a guy who talked about multimedia in a DITA-based CCMS and the tekom CEO, who introduced the in-progress standard for multi-format content creation. Some of the writers in the audience were getting frustrated with all the tech talk. One asked, "Isn't it enough to be a good writer?"

My answer at the time was, "If that is all you can do, you'd better be the BEST writer in the world. You'd better type fast, deliver on time, with absolutely no mistakes."

The truth is, not only is that not enough, it's not even the same profession. A person who has perfect grammar and high productivity, and not much else, will be hired to be a knowledge base editor. If you find an employer willing to pay more than minimum just to edit Support articles, keep your CV updated. Sooner or later, your employer will realize he cannot afford you. The initiative will be dropped, or the position will be outsourced to a cheaper supplier.

Remember: a love of learning and a fascination with technology are characteristics of a great tech writer.

7.7.2. Innovating According to Plan

Before you begin innovation, make sure you have a vision. If you engage in new team projects without a guiding vision, the chance of a successful end result is random, at best. Then, make a plan for End-to-End innovations.

1. Make a commitment to innovation.
    First, make it for yourself. Look at your schedule and the company's roadmap. Where
    does innovation fit? If you cannot find an immediate place, re-prioritize the products,
    their features, or the features of your deliverables. What can be postponed or dropped,
    and still be good enough? Connect innovation to your vision, which connects new
    ideas, methods, and technology to the future of your team and pulls the commitment
    along a natural path.

2. Communicate the commitment to your team.
    Ask for their input. What do they want to change? What experience of successes or
    lessons learned do they have? Let your team brainstorm and share. This is an
    opportunity to raise motivation through engagement, appreciation, and creativity. Take
    advantage of the chance by listening, accepting ideas without judgment, and being
    open to learning new ideas.

3. Give your team a template that guides them through the engineering process.
    Make their ideas part of their required expectations, and make the submit-approval-
    implement process clear.

4. Schedule time for innovations for each person on your team.
    If some employees do not submit feasible ideas, they can help with the process and
    implementation of their peers.

7.7.3. Innovating with New Technology

The tools that tech writers use to create content are developed and updated all the time.
Some of these tools are specifically for authoring, and others are not.

A prerequisite to every implementation of innovation is ROI: Return On Investment. A
clear ROI means the ability to prove a future profit from an investment with reduced time,
less headcount, and less annual spending. Before you start looking for a new thing or idea,
make sure you have productivity and deliverable stats, to be able to show your budget-
controllers the before-and-after graphs.

When to Bring a New Authoring Tool of a Known Type

If you bring a new tool of the same type that your team already uses, make sure that you
and your team buy in before you implement. You must all understand the benefits of the
new tool, engage in the learning process, and commit as stakeholders to the change and
deliverables.

Bring this type of innovation only if you are convinced of the ROI.

When to Bring a New Authoring Tool of a New Type

Innovation can be the acquisition of a new tool of a different type. For example: replace a
file-based tool that delivers mainly PDFs (Microsoft Office, Adobe Framemaker, Google

Docs) with a chunk-based CCMS (Component Content Management System) that delivers single-sourced multiple outputs (AuthorIt, Flare, Paligo, proprietary database for XML topics). The challenge of achieving 100% buy-in of the team is greater. Added to this is the challenge of implementing the tool without suffering the heart-clenching fear that comes with not knowing what you are doing.

Bring this type of innovation when:

- The ROI is clear.
- The new tool is able to deliver innovative output that fits your target audience.
- You have your team's support.
- You can work overtime for a while.
- You have a budget for a consultant, or your vendor gives free training and setup.

When to Bring a New Non-Authoring Tool

Technologies for R&D, QA, and Support can be priceless for tech writers, when implemented correctly. Internal wikis can help you plan, maintain a dynamic and useful style guide, and motivate your team by giving each individual a voice. Controlled access to the code, with understanding and parsing tools, can solve the issue of uncommunicated changes and enable one-to-one translation in docs. Access to QA test plans with procedure-led collaboration can make QA and documentation more efficient. Access to Support tools (such as Zendesk or Guru) can raise the quality of articles and alert the writers to issues that should be documented. Installation of script compilers can raise writing productivity with automation.

Bring this type of innovation when:

- Your writers are technical enough not to make stupid mistakes, such as deleting months of work.
- Your writers have work-hour bandwidth for new responsibilities.
- Your writers see these new challenges as opportunities to improve themselves or to engage in tasks that are more interesting.

7.7.4. Innovating with New Methods

"Those who initiate change will have a better opportunity to manage the change that is inevitable."

—William Pollard

Effective companies innovate frequently. They measure productivity and look for areas to improve. Make sure your team is not constantly on the cast-off list! Manage your own productivity. Find ways to measure your efforts, and analyze the results to clean house, before it is cleaned for you. Keep up with the changes going on around you. For example, if VPs start to send doc feedback, find out if upper management has new duties to review docs. If so, proactively change your review procedures to include these managers directly.

The challenge with changing methods of technical writers is that most writers prefer stable procedures. They know what they are doing now, and they know what to do next. When you change that, you rock their world, and not in a good way.

Innovate method changes when:

- You carefully follow a change management plan that includes the individual needs and quirks of your employees.

- You have your manager's support.

- New deliverables are tested, for security and for feasibility.

- You can measure and follow-up on issues to tweak.

- Your team has documentation of the methods and can follow a written procedure until it becomes natural.

Make sure that you implement an engineering process, with effective communication.

Innovating According to the Engineering Process

The engineering process is well developed. If you do not like my steps, create your own. But give your innovation a chance to survive birth: use a process.

Example of engineering process in documentation innovation:

1. Estimate the complete effort of this process, and enter it in the participants' schedules. The best way to kill innovation is to assume it can be done on downtime. It is often less *How long will this take?* than *How much time can I allow each person on the team to work on innovations?*

2. Define the issue to be solved.
    Do not change for change's sake alone. Define the success criteria of a solution. Analyze the audience who will use the innovation. Make sure it is something that will be implemented.

3. Research known solutions.
    Do not reinvent the wheel. Learn from the trials of others. Find out if there is a tool that already exists that will give your team the same solution.

4. Choose a solution, in writing, with documentation of its prerequisites and risks.

5. Bring in your manager.

   Before you develop the new methods, make sure your manager supports you. Your manager might also have knowledge that will change your solution. Maybe they have prior experience or know of another team working on something similar. It can really piss off a time-stressed manager if you go beyond this point without consulting them. And your manager will be your project champion, defending your innovation when others will more authority and less foresight would try to kill it.

6. Design the solution.

   Bring in one or two of your writers. If the new procedure is created by a committee, your team will be able to accept it more easily. During this phase, continuously analyze the solution: Can you mitigate or avoid the risks of the design? Can you make the solution simpler?

7. Develop or acquire the solution.

   Decide if you can develop it and if you should. If you need a new script, maybe you can ask an R&D or QA team manager if they can do it for you. Even if you have the knowledge, you might not be the best person to do the job.

   If you are not creating the solution yourself, stay on top of it. Make sure it is as simple as possible, includes the success criteria, and avoids or mitigates known risks.

8. Test.

   The design committee (subset of your team) can run the first tests. Look for steps that can be simplified or eliminated. The more complex a procedure is, even if each step is simple, the less your team is going to like it.

   Continue the tests with a wider pool. Bring in other writers or let QA give feedback on the deliverables.

   It is important that you create the pilot or prototype as early as possible. This phase should take the longest. If it takes 8 hours to reach step 6, and 16 hours to complete development, the testing phase should require at least 24 hours.

   Make simple test plans, such as a list of checkboxes or sample scenarios. Then, divide the hours by different people. You can complete 24 working hours in an 8-hour working day if you assign 3 people to it.

   If the solution fails, decide if the failure is a show-stopper, a lesson learned to guide changes, or an incorrect test.

9. Analyze the solution.

   If the solution meets the success criteria, communicate the new change carefully.

   If it does not, document lessons learned and go back to the first step. If your first definition of the issue is not correct or complete, start over. If the problem is somewhere else in the process, start at the appropriate step.

10. Follow up.

When the innovation is implemented, analyze its actual success and areas of improvement. Write a success report for your manager and your team, or schedule time to design, test, and implement fixes.

If nothing moves:

- Send an email: "I'd like to discuss…" or if waiting for action item: "When can we expect…?"

- Set a meeting with the goal of getting a milestone commitment. And then follw up until closed.

7.7.5. Change Management

Implementation of innovations requires *change management*. Change management is a skill. It is a science: there are givens and procedures proven by experiments. And change management is an art: it demands talents that can be improved with practice.

Use change management when:

- You bring innovations to the team or company.
- Your team or organization changes people or hierarchy.
- You bring a known technology or method to a new audience.
- You bring back old ideas.

With a PAEI[13] model analysis, I was an *Entrepreneur* type, and most of my team were *Administrator-Integrator* types. Suddenly my worst frustration was made clear. The ideas that banged into my head and made complete sense to me, were not being communicated well to my employees. The benefits were not intuitive to them. The implementation that seemed simple to me looked like overhead to them. I learned to communicate my ideas, no matter how simple they seemed to me, as changes for them. I learned to manage innovations with a change management model.

Prepare for the Reactions

Predict who will have a negative emotional reaction, who handles change badly, from past responses. Most technical writers thrive on order, structure, lists, and procedures. When you ask them to change what they do, some will have negative reactions.

How will this change affect their values? List possible value conflicts. Prepare keywords to alleviate or prevent the conflicts and to align with their values.

For example, you decide that your team will divide the tasks of editing a large section of the Knowledge Base. Each person will do one a week. A typical technical writer response

---

[13]The PAEI model was developed by Dr. Ichak Adizes, management expert and founder of the Adizes Institute.

is: "I do not multi-task. Every time I take on a short project, it breaks the flow of my priority projects. I am not a robot." Now you can decide if the once-a-week idea is a good one, or to change that to be a monthly project. You can build a response that will help the writers understand the importance of this issue and how to move from one project to another more easily.

Know your audience: your team and your manager. Do not bother your manager with changes to team **methods**, unless you are sure the manager is interested. Do not implement a change to **deliverables** without your manager's approval. Some changes are relevant for your users. Define your audience and then analyze their possible reactions.

- How does your audience perceive what you have now?
  Do the writers see their current methods as cumbersome? Will the change ease that or make it worse? Does your manager see your productivity as low? Will this change help, or should you mitigate with ROI explanations? Is there a probability that user satisfaction will be affected?

- Is the change going to be considered new?
  Be aware of the human fear of the unknown. Be prepared to answer questions that are based on this fear. Give your team time to process the solution and get over the fear.

- How much learning is required by you, your team, and your users?
  Is someone going to worry that they cannot learn as fast as the others? Are they afraid they will lose status or responsibility if they cannot keep up? Consider these common fears, and find a way to alleviate them.

- Are behavioral changes required?

- Is this new to the tech writing field and competitor deliverables?

- Can you sell this change easily?

Gather Your Allies

Before you communicate the change to the full forum, have a chat with one or two people on your team. Choose the employees that you think will take on the change most easily or the employees who influence others with positive energy. These employees will be your litmus paper for objective implementation of the message.

Choose a friendly time and place. Find these employees getting coffee, step into their office to say hi, or anywhere that isn't you sitting behind your desk. In a by-the-way manner, get their reaction to the idea of the change.

As you talk with these employees, try to recruit them to your vision, and accept their responses. Listen to their reservations. With pre-knowledge, they will be your supporters when you bring it up in the team forum

Communicate Your Passion

When you are ready to communicate your innovation to the team, use your passion.

Explain the change with conviction. Start from a point of their *What's In It For Me* questions. Give them a meaningful understanding of the need for the change. Give all the information that you can give, without going too deeply yet into difficulties or technical details. Connect goals to future targets. Build a picture of success.

Be prepared to fight the small-mindedness and negativity with positive energy. Throughout the process of the change, reiterate its long-term benefits. Let your people vent their fears. With patience and understanding, validate their emotions and then find solutions. End with another repeat of the benefits. Be a broken record reciting the purpose.

Have frequent informal talks in the beginning. Plan for more time than usual for such informal talks. Enforce one-on-one, team meetings, and other formal communication with people involved in the change.

Be there for your people. Be around physically. Move around and talk. Be available to answer questions. If you don't have an answer, don't give something off-the-cuff. Say you will find out, investigate, and come back with an answer.

Be a personal example. The management of your message is as important as your words. Bring work, numbers, and data with your message.

Come with a Plan

Consult with your manager and with allies on the team. Set up training and knowledge sharing. Test the idea and run a pilot. Document the results.

For each stage, communicate with actions and deadlines. Be clear about your expectations. Assign roles for your employees to run their own pilots.

In your mind, commit to basing the go-ahead on their results. If the idea does not save time or provide better results, or somehow give everyone a real benefit, you must drop it. When that is clear in your own mind, let your team know that you are committed to their results.

But at the same time, be aware of push-back that will come with the results. You must review the results. Sometimes your people will delay change by putting off their roles. Set deadlines. Often, they will find in the results what they expected to find:

"We tried. It didn't work. It's too expensive. It's too big. It's too complicated. It isn't practical. We don't have time. The team didn't accept it. Let's think about it another time."

If you are sure these results are incorrect, do not give up! Discuss with the naysayers and go through their pilots and results in a two-way dialog.

Follow Up

Make sure everyone has what they need to face the change and implement the solution. Do you need to give more time to individuals for training?

Get everyone back to work as quickly as possible. Enforce the routine. Nurture an environment of doing. Celebrate success and show appreciation for results.

7.8. Handling Conflicts

Handle conflicts quickly. Protect your employees correctly when there is a conflict with another team.

There are a lot of books and courses on conflict management. Most of them are created by people with more experience or more knowledge than me. In this section, I take the basics of conflict management and tweak it for your everyday needs as a manager of technical writers.

One type of conflict you will find is between your writers. One wants a project that was given to another. One complains about the character traits of his officemate. Another is about to explode with frustration over the nasty peer review remarks. All of these issues can be dealt with standard conflict management. The important notes here are to 1) actually deal with the conflict and 2) be consistent, fair, and trustworthy (follow through with whatever you decide).

A different type of conflict is unique to technical writers. Your writers have to work with people from all the other departments. For any deliverable, they must cooperate with Legal for the copyright, with Marketing for the overview, with Product Management for the priority of features, with QA for technical review, with R&D for fundamental understanding, and so on. When one of these contacts has a conflict with your writer, you must manage the conflict for the best result for the company and for your writer.

7.8.1. Identify Conflict and Create a Plan

1. Decide that you have a conflict.
   There is a contrast of goals between parties with conditions that block mutual satisfaction.
   If a project manager wants to vent on one of your writers for not having last-minute GUI changes reflected in the document quickly enough, her want is her goal. It must contrast with your writer's goal. No one likes to be yelled at.

2. Define the goals of all sides.

3. Define the relationships, resources, and acceptable costs.

4. Choose a method. Define your options - together or alone.

5. Implement the selected option.

7.8.2. Mix the Wines

There is a Jewish fable of a king who was given two cups of wine.

"One cup is the wine of Mercy. When you drink it, your rulings will be filled with mercy," he was told. "The other cup is the wine of Justice. When you drink it, your rulings will be just."

When the king held his first judgment court, he drank of the cup of justice. Every thief was led to the ax, whether he stole gold for greed or bread for his family. The people feared the king.

When the king held his second court, he drank of the cup of mercy. He gave everyone a second chance, speaking kindly to all. The people said he was weak, and criminals laughed at him.

Before the third court, the king held the two cups in his hands. Then he poured some of one into the other, and back again, until the two wines were completely mixed. The king gave mercy to those deserving mercy and strict justice to those who deserved such.

A good manager is assertive and understanding. Neither is a negative characteristic. Neither promises great leadership. Both must be used in balance and correctly for each situation. With the conviction of principle and empathy for your employees, you will identify conflicts before they become problems. You will have the willpower to confront the issue in time to gather relevant information and create a plan of action.

If you are assertive without caring about the people, your solution to conflicts will often result in competitions. Competition is not negative in itself. But under-managed competition can create a negative, hostile work environment. If you lack empathy, you will not be able to prioritize conflict management. If the conflict is between your employee and a manager of your level on a different team, you will not handle the conflict, or you will tell your employee to handle it. The first causes a growing rift between your writer and the SMEs, which can make it impossible for the writer to complete tasks on time. The second will cause a rift between you and your employee ("my manager does not protect me") and between you and your peer manager ("why am I am doing my peer's job?").

If you are empathetic but unwilling to use your authority, your solution will be surrender. This will result in inconsistent management. Your employees will lose trust in you and in your ability to manage them. Sometimes you will ask one side or the other to surrender. The conflicts will fester or explode. If you ask one person to surrender too often, that employee will most likely leave your team.

Your first choice of conflict resolution method should be cooperation – a mix of the wines that depends on all sides being willing and able to take an active part in the solution.

Cooperation takes time, so consider carefully. If the conflict is too trivial, using a cooperative methodology will make a mountain out of a molehill.

7.8.3. Example of a Conflict with a Writer

"You look pensive."

"I have a one-on-one with Alex today. I expect a conflict. I'm handling it, but this is a heads up."

Prepare your manager without defending your goals. If the employee escalates the issue, the manager is not caught off-guard but can manage the conflict objectively.

"I'm not done with feature X because you keep rejecting it. Can you please just look at it now and approve it?"

"I rejected it because you continue to write in passive. I told you to change the highlighted sentences to active. You rewrote them, but they are still passive."

"I'm not as good with grammar as you. If you want it written differently, just write it yourself already!"

"I'm really disappointed in your response."

Do not show weakness! Manage this conflict. Do not avoid it.

"If you are unsure of the difference between passive and active, there are sites that explain it. Alex, you are a professional writer. You must gain an excellent understanding of grammar. You must be able to write according to our style guide."

"What about this task? It is due today! Where am I going to find the time to learn grammar and complete this on time?"

"OK, let's compromise. Get your laptop and come back. I will clear my schedule for the next hour. I will explain passive versus active, and you will change your writing for this task with me next to you. But Alex, you must take the time and make the effort to understand this. You must prove to me during this time that you are learning this and changing your writing habits."

7.8.4. Example of a Conflict Between Writers

"I just can't work with Betty anymore. Ever since you made her technology leader, she's been trying to tell me what to do."

"You and Betty have been working together for a long time. Let me see what we can do.

"Betty, can you explain the situation with Alex?"

"It's a problem. I gave him the data from the servers, but he won't write up the server procedures."

Later...

"Betty, Alex, thanks for coming to talk. We have this project that includes work in the server room. Alex has a project plan. Betty, you have server data. Let's discuss how to best cooperate on this."

"We don't need to cooperate. She just needs to give me the data and let me handle it, if I decide it should be documented."

Betty: "Now wait--!"

"Let me interrupt here. We need to recognize that you both have a stake in this. We're going to discuss this. And we will reach a conclusion. Listen to each other. It's worth the few minutes out of your day to settle this."

(As a mediator between your employees, set up the rules of the discussion. Help them see each other's point of view.)

"Betty, do you agree that this is Alex's project?"

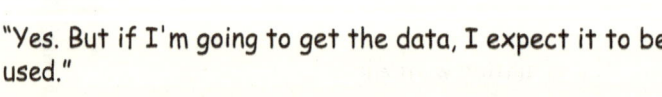

"Yes. But if I'm going to get the data, I expect it to be used."

"But we're not at the point in the plan to use that data. And when we get there, it will have changed. You'll have to get it again."

"Hm...maybe we need a workflow? The lead writer shares the plan with the tech leader. The tech leader gets the data according to the plan's schedule. What do you think?"

"That makes sense. Alex, I'm sorry I lost my patience with you."

"That's OK. Thanks. I'll add that workflow to the plan and to the procedures for this type of project."

7.8.5. Example of a Conflict Between a Writer and an SME

"I have a problem with the manager of Area X. When he started the project kick-off, he said he wanted to review all the documentation himself. I have sent him the draft and several reminders. He said he didn't have time. I suggested that he delegate the review, and he got quite nasty. He said I should learn the product myself and stop bothering him with trivial things like the books that no one reads."

"This is really important, Betty. I know you've had a hard time getting reviews, but I didn't know you were being disrespected. That is unacceptable. I will talk to him."

"Oh! No! Please don't! I just wanted you to know what's going on. I don't want to make it worse."

"We don't want bad feelings or back-stabbing in the company. We have to talk about it. Would you rather you spoke about it with the manager yourself? I can give you some pointers."

"He won't meet with me anymore. He's really stressed."

"One of us, you or I, must speak with him. Look at how this is done. First, we identify that we have a conflict. It's not just an argument. You want something – to complete the docs - and he wants something else – to complete the product itself. This is his first big project. He wants to control it all. He wants to prove himself. We have to think of his position, and we have to get our goal across to him, as something he wants too, in a way that lets him save face."

"When you spoke to him about delegating the review, was he alone?"

"No. He shares an office with the other manager. But he wouldn't go to a meeting room. He said he was too busy."

"This is one of the advantages of letting me talk to him. He shouldn't refuse to meet with a team leader. If necessary, I can escalate to his manager, Rani. Betty, it's for the good of the company, not just you or our team, that we address this issue. I promise, we'll find a solution."

Later...

"Thank you for meeting with me. This won't take long. I understand you want to drop the documentation from your project definition."

"That is not true! I must have the docs! And they should have been done a month ago and delivered to the translators."

"That is what I thought. We actually did send the documents, without your review, to the translators. You were on the CC of our budget and status report for that. You know, that is in conflict with the methodology. If you want major changes after the translation, getting the fixes in three languages could put your project over-budget. The problem now is that you told the writer you didn't care about the documents. We have a procedure that includes R&D review. You know that the writer learned the product in the beginning. She wrote a very good document, as far as we can tell. But only you – or your team leaders if you choose to delegate the review - can say if it meets your requirements."

"I don't have time to read the book. I have more important things to do."

"Thank you for meeting with me, Rani. We have a conflict with Robert that I'd like to see resolved."

"I'm not surprised. He's very stressed because the widget doesn't fit the wadget, but we still must deliver to an important customer on time."

"I'm glad there is a good reason. We want to help, not hurt the timeline. We can make an exception for the review process this time. But we still have an issue of how he spoke to my employee. He was disrespectful, and with another person in the room. If that isn't resolved, I will have to move a different writer to complete the project. I don't think that will do good to the timeline. And I think that the

current writer is doing a great job. We'd all like to continue on this, but with better relationships."

"You are right. I will speak to him."

A few days later...

"Betty, how is it going with Robert?"

"He actually apologized to me. And then he sent me a list of names of people in his area, to send the docs for review."

7.8.6. Example of a Conflict with a Peer

When in conflict with someone on your level, bring in your manager. Do not try to mediate where you cannot be objective. Do not step out of your domain without backup.

"I have a problem with the Training team leader. I set up a meeting and invited you. Please come."

"Sure. What's up?"

"We agreed months ago that we would share the load of setting up environments for hands-on. Now he's saying they don't have time and demanding we do all the environments for both teams. "

"I see from your schedule that we are tight too. You gave your whole team one day to set up environments. If we double that, we will be late on three projects. I will help you resolve this."

7.9. Accepting Accountability

Bad things happen. Sometimes they are mistakes that we make. Sometimes they are the product of a flawed environment. Sometimes they are mistakes of others that affect our own output.

Responsibility means being open about the mistakes we make, and doing what we can to fix them. Accountability means you take responsibility for all issues you discover, regardless of origin, and you follow through with solutions. You give up responsibility for blaming yourself or others, and take on accountability for the solution.

At first, writing a report of an issue can be seen as a time-consuming, unpleasant task. But you will soon see that the main purpose of the report is to find the silver-lining in a horrible mistake or oversight.

7.9.1. Accountability Flow

1. Problem discovered.
    Does the corporation see it as an issue or as something to be leveraged? Be aware. Politics are always present. If others disagree that this is an issue, fight only to promote your agenda as the customer advocate or as the team leader. Continue only if a solution to the problem will help the users or will help your team help the users.

2. In an Issue Report template, define the problem clearly and simply.
    Why is it a problem? Who does it affect? How does it affect our end goal?

3. Find the cause of the problem.
    Is the origin outside our team? If so, what can we do or change to help others prevent the issue in the future? Do not define the cause in terms of blame. Define it in terms of what can be done better.
    For example, "The SME refused to tell us when the GUI changed" is less likely to lead to a solution than "There was no process for the SME to communicate GUI changes to us."

4. Suggest a solution for this issue, now.
    If you think there is nothing to be done, look again. This is probably the start of a real innovation that you can lead.

5. Suggest prevention. Better, suggest a few preventative solutions.
    Be careful about solutions that require changes in other teams. You will be leading the solution, and it is frustrating to have the responsibility without the authority.

6. In a scheduled meeting with your manager, ask for their review and approval of the Issue Report.

7. In a scheduled meeting with the stakeholders of the issue and of the solutions, create a plan of action.

   This includes the names of people responsible for design, prototype, implementation, and test of the solution; the schedule; and the priority.

8. Follow up.

   Make sure the implementation is on schedule according to the plan. Make sure that the solution fills the need. If it doesn't work, start over.

7.9.2. Issue Report Follow Up

Someone on your team takes accountability seriously, discovers an issue, and sends you a report. Now what do you do? Your next steps or inaction will lead the culture of your team. Is accountability rewarded or punished with more tedious work? Do you follow through? Do you protect your team when others cast blame? Do you push for what your team feels is important when others want to drop the issue?

Guidelines for follow up:

1. When you get the first report, suggest fixes to make it shorter and more about the solution than the problem.

2. When you and your employee agree on the report, help them make a list of stakeholders.

   Who cares about this issue? Who can work on the solution? Who can use the solution? Help your employee communicate with the stakeholders and get their feedback. Help them accept it bravely and without defense that can put off the desire to send more feedback.

3. When the stakeholders agree on a solution, help the employee write an Action Items table: tasks, owners, and target dates.

4. If the employee owns action items for the solution, help them change their schedule and weekly goals to fit in time for these actions.

   Ask them to keep notes of their part in the solution, for their performance review, or find another way to reward them for this extra work.

5. If the action items are on other people, decide if the writer can still own the solution. They will have to communicate the actions, follow up with status reports, and make sure the solution is implemented. If the action owners are higher up on the hierarchy, it might not be feasible for the writer to own the solution. Do not give responsibility where lack of authority makes follow up awkward or impossible. If the writer cannot own the solution, make sure to give the writer credit in the end communication and wherever possible during the actions.

6. Follow the status of actions based on the due dates given.

When I write an Action Items table, and get commitment from the owners, I enter a reminder in my schedule to follow up on each due date. If the writer is the owner, help them do the same. If an action deadline is missed, update the dependent due dates, and send the update to all the stakeholders.

7. When all actions are done, test the solution, communicate the success, and thank everyone for their efforts.

   If the solution is a change in methods or processes of your team, take responsibility for making sure the new way of working is put into practice correctly, immediately, and consistently.

7.9.3. Issue Report Template

Keep the first report template simple. I suggest that you keep the report on one document or wiki page. Link to the plan and follow up resources, that you keep on other pages. You can have an Issue spreadsheet, with the status and links to all Issue Reports.

Headline	*Short (less than 100 characters) descriptive title and the subject of all emails of this issue and the solution.*
Date	*date submitted*
Reporter	*Name and username*
Description	*Full description of the issue, in terms of its impact and severity, but short and to the point.*
Cause	*Short explanation of the cause, in terms of missing or incorrect procedures or tools, not in terms of blame.*
Fix	*Solution for the issue to be fixed immediately.*
Prevention	*Solution for the issue to be avoided: the change or new technology or procedure to be implemented. List of people to whom this is communicated.*

7.10. Leading Technical Writers

"To handle yourself, use your head; to handle others, use your heart."

—Eleanor Roosevelt

"The first responsibility of a leader is to define reality. The last is to say thank you. In between, the leader is a servant."

—Max DePree

You can manage a team with priorities and plans. You can manage your own projects. But a master of human alchemy leads people. You lead them to self-improvement, skill improvement, and self-management. Leading technical writers is a niche with specific issues of its own.

What comes next is from my professional journal. Most I learned from trial and error. Some I learned from management or leadership workshops.

7.10.1. Leading Proactively

Do not wait for issues to come up. Find the questions that users will ask before they ask. Find solutions before issues become problems. When someone on your team finds a questionable bit or a possibility for improvement, lead them to own a proactive solution.

If you are proactive, you and your team will have the opportunity to have an impact on the areas you believe are important.

Being proactive requires a passion for quality and the disciple to follow through. It takes time and stamina to schedule a project that is not requested by others.

1. Identify the existing need that can be met or future problem that can avoided. Make sure the solution will have a positive impact on performance or results. It must be worth the resources.

2. Create an action item or complete project, according to the scope of the issue. Set clear boundaries.

3. Set a relevant schedule.

4. Implement the plan.

5. Review the results.

7.10.2. It was a Team Effort

Encourage team cooperation. For example, you can set a lead writer and a second writer for every project. Every writer has their own projects, but they also stay up to date on their partner's projects. This helps with peer review enforcement, informed proofreading, emergency hand-offs, and documentation sprints. It can also bring a little healthy competition that leads to increased quality. When a writer knows they are working with another expert, they will do better on a daily basis.

It can also be the foundation for a great professional relationship that fulfills personal needs. The two writers become sounding boards for each other. They deflect emotional reactions before frustration reaches an SME with disastrous results. Your team will find more pleasure in coming to work. The personal bonding is like any good relationship, somewhat euphoric and almost addictive. Imagine your employees being addicted to work!

Now, go back to "can," the second word of the previous paragraph. The type of relationships that are created and reinforced in your team depends on your management skills. If your team environment is filled with negative energy, the enforced partnership will be ammunition for mutiny. If you gossip about one person's weaknesses with others on the team, if you vent to others and it gets back to the team, if you degrade or punish an employee in front of others, you effectively poisoned the team's cooperative efforts. This is more than simple respect. This is you, understanding that the efforts of your team are your responsibility and that these efforts depend on human beings who need respect, security, fulfillment, positive energy, and everything else that comes from you shutting your mouth when it is correct to do so.

Productivity and quality are symptoms of human energy.

I can't find a source for this quote. It must be mine.

"You want employees energized enough so they perform productively and put in extra effort. But push them too hard and risk the onset of burnout, culminating in absenteeism."[14]

7.10.3. Recognizing and Communicating Benefits of Change

Whether a change is actually a good thing or not, this is your mantra: "Change is a good thing." When you can, prepare for changes from inside and outside the company. When you cannot prepare, embrace the change. Make it a good thing. Find the third door and follow through.

Be willing to take risks. This is a sign of a mental strength and self confidence. You can change because you are a strong person. You can lead change in your team because you are

---

[141] Adonis, James. http://www.jamesadonis.com/Newsletter_Strategy_and_Direction.htm

a leader. You are willing to go out of your comfort zone because you know that discomfort is the best path to improvement. You are willing to fail because you know that failure is a more valuable learning tool than success.

7.10.4. Integrity and Ethics for the Writing Team

If a writer is performing badly, and you do not tell them (or worse, you tell them they are doing great), your lack of integrity will come to bite you in the behind. First, you rob that person of the chance to improve. Second, you can find yourself in a legal quagmire if you decide to let that person go.

Be open about your own mistakes. If you make a mistake, act as you wish your employees to act. Admit it, in writing if the mistake affected output. Find a solution. Apologize if appropriate. In return, when your employees are open about mistakes, act as you wish them to act when the tables are turned. Forgive, take it professionally and not personally, and work together to define the lessons learned. If your team makes a mistake, take responsibility. Upper management doesn't want to hear names. All they care about is what you are going to do to fix it.

7.10.5. Knowing When to Shut up

Technical writers in particular are, or feel they are, less connected to the culture and current events of the company. Most of the people around them are deeply involved in their team or group. The technical writers are more likely to know more people of all of the groups, with a more shallow connection to the issues of one group but more opportunities to overhear mentions of new projects or events. Thus, the writers often feel they are missing information.

Do not answer direct questions for which you did not prepare. It is better to refuse to answer or to say you will get back with the answer. "Why isn't Alex on this project?" "Why do we have to move rooms again?" "Why did that project manager leave the company?" An off-the-cuff answer may turn out to be inaccurate or become a derogatory statement about the company or a person. Such statements will degrade the respect in which your employees hold you and build a muddy foundation of fear of what you are saying about each of them when they are not in the room.

Some information is withheld by company policy. If a big change is coming, you might be informed early, to be able to assess its impact on your employees and help the decision makers decide how best to communicate the change. If you let this information slip before you have approval to speak of it, you ruin the effective internal communication initiative. This directly leads to lower morale, lower productivity, and higher turn-around. It is not about keeping secrets from your employees. It is about letting the company deliver information in the best possible manner to be fair to all.

Do not communicate issues to your team until you have bought into the company's stance, or you can stand behind the company's decision as though you agree, even if you do not. You must take responsibility for the decisions you deliver, even if they are unpopular.

7.10.6. Improving Your Patience

You cannot control all the external situations, but you can control how you react to them, without changing who you are.

When you are in one-on-one meetings, work towards your end goal. If you do not have a goal for your employee or manager, your job in the meeting is to listen. If you are not absolutely sure about any issue that comes up, say, "I'll have to investigate. I'll get back to you on that." If an idea or complaint comes up that you disagree with, ask for more information. Do not react emotionally.

7.10.7. Encouraging an Effective Environment

If you find that employees knew about a problem before your manager told you, you have a change to make. Why aren't your people talking to you?

- When you get negative feedback, do you defend yourself? Or do you avoid getting feedback at all?

- Do you put your trust in upper management as a mechanism to avoid taking risks?

- Are you unable to schedule time to take real action?

- Do you show your employees that you don't trust them?

- Do you avoid conflict and diverse opinions?

If these behavior descriptions are true to you, you created an environment of silence in your team.

The change to make is in yourself. Encourage employees to talk. Find value in what they have to share. Each person's contribution has the potential to influence decisions. It is your responsibility to gather all the offerings, and identify the potential in the ones to use. You can do this if you commit to hearing your people. With genuine interest and sincere trust, ask your employees for their opinions, their observations, and their fears. Show them you are open to change. Prove your tendency to take action and your willingness to face conflict.

7.10.8. Identifying Over-Protection

At first, I had something trite here: "Ask of your employees only what you are willing to do yourself." Well, I do believe that if you do not have the required knowledge or skill for a new output, you ought to learn at least the basics yourself. But the important thing is that you are an example of the sort of human you expect your employees to be. Do not be an example of an exceptional writer – you are not a writing resource!

Here is an example, and appreciate it because it's embarrassing for me to relate.

I had a team that published 20 documents exactly on the hour that the product was released. We set up the two different outputs, ready to go, on the day of the Go-No-Go. But someone had to be there when the product was declared a Go and uploaded to the Internet. At that moment, the technical writers joined two other teams in the final release.

I thought the whole synchronized release was exhilarating. We all worked together making sure it went out just so, right on time. It was even more fun when it there was a No-Go and a call to wait. Developers would come in and fix the last thing, or a build would be remade, and the synchronized release would happen at midnight. Being a part of the stress and bonding was something I looked forward to.

Then it came to a day when I couldn't be there. I wrote a procedure for the team and told them to handle it. Oh my goodness! I should really have predicted the negative response. No one was interested in staying late for this. It was too late to ask for VPN access for the team to do it from home. And the worst result: when I wanted to promote someone from the team to be team leader, he said he didn't want the job because he couldn't put in the hours I did.

I'm so ashamed. I'll tell you why. I could have said, "This task belongs to the team. We will do this together. You can be in the office or you can get VPN access. I will lead the logistics and divide the uploads. You will each upload and switch to public when I say Go. On the day of a release, you must all be available, at home or in the office, until it's done. This can be until midnight or one AM. It doesn't happen more than once a quarter, so I expect you all to do what you can to make this happen smoothly and correctly."

If I had done that, everyone on the team would have been a part of the company's smiles the next day. They would have learned the skills necessary for a synchronized release. They would have been more a part of the company and would have had a chance to shine as a team. I could have offered gratitude prizes, such as a half-day off the next day or a big team breakfast.

Instead, I changed the definition of technical publications team leader to include tasks that I should have delegated. I lost a potential manager. My team lost the chance to improve and shine.

Being the example does not mean staying later if a writer must stay late. It doesn't mean coming in on a weekend if a writer must. It certainly does not mean being a great writer and taking over writer responsibilities.

Being an example means you constantly improve your soft skills. You are careful of your tone. You speak with tact. You encourage goodwill in all you say and write and do. You bring positive energy. You take care of your health and prioritize your life. You are focused on the task at hand and keep the big picture in your mind at all times. And unless you are more of a saint than me, you give yourself constant fair and honest feedback, and continue to improve yourself.

If you over-protect your team, you all lose out.

7.10.9. Flipping Issues

Do not blame anyone, even in private. Look at each setback as an opportunity. It sounds cliche, but it really can be done. All you need is to close your door for a blocked off hour. Consider the situation without emotion. Drop your anger and frustration. Forget about the clock. Focus on how to turn this problem into a solution. You will come out of your shell filled with gratitude and positive energy to get this solution started.

When you must deliver bad news or make an unpleasant request, own it. Never lay it at the door of your manager or the company. "The VP said all the tables have to be aligned," is going to make more problems than, "We must make a macro to align all the tables in the docs, and we have to republish the docs for the version we just released." If they ask why, the best answer is, "We have feedback about this, so I consider it a high priority."

You are not your employees' friend. You are their mother...um...manager. You must make decisions that are for the best of the team and the company, even if someone else will hate you for it. Refocus their "hatred" to yourself, and accept it.

7.10.10. Maintaining Energy

Be aware of your own energy. Ask yourself, is it because of me that my team is stressed and bored? When you are low, make sure you are taking care of yourself: eat healthily, sleep well, exercise, and take breaks. Make sure you are not complaining or gossiping yourself.

Let your employees know you believe in them. When you give a new goal, end with, "You can do this." When you give negative feedback, make it an opportunity for them to take control of their own improvement and solutions. When you explain your expectations, make it clear you know they can meet them.

Find the employees who have constant higher energy levels and show your appreciation. Be candid, and ask them to use that positive energy to influence their officemates and teammates as much as they can. Find the employees with weaker energy and let them know their cynical comments and jokes have an influence on others, and that you believe they can become sources of positive energy if they choose.

If nothing else works, break the routine. Take your team out for brunch. Use your team budget for a small event. Redesign project methods to cut out or spread out tedious tasks. Hold a guided brainstorming session.

7.10.11. Showing Respect

These are professional adults. They earned their place on your team, and they deserve to be treated well. Trust their intelligence and encourage their drive to be part of a successful company. Make decisions based on the fact that they are equal human beings, each with different skills and strengths, and each deserving of opportunities to improve and to shine.

Encourage and respect different opinions. I once was in another manager's team meeting. First, he introduced me to the team. There was a writer whose parents emigrated from India, and a writer from Texas. The graphic artist was a new immigrant from Slovenia. The tech guy who handled scripts was on a work visa from South Africa. During the meeting, one of them suggested that a procedure be changed. The manager interrupted and shot it down before she was done explaining her idea. That group was a color wheel, not a team that encourages diversity.

7.10.12. Celebrating Success

Show your appreciation of your team's efforts, but proportionately to the effort extended. If someone does a great job on an important project, don't make a bigger show of appreciation than for another person who did a great job on a less important project. You assigned the projects, so to be fair, base appreciation on effort and results.

Expect your employees to shine with excellence, and reward their results. Do not discourage employees from surpassing you. If you are worried about an employee becoming your manager, improve yourself.

7.10.13. Keeping Your Promises

If you say one thing ("You will get Project X,") and let another scenario happen ("Sorry, Charlie. Alex is taking Project X,") your writers will learn to not trust you. You will find yourself having to repeat instructions and updates. Your writers understood you well the first time, but they don't believe you. Later, you will find yourself talking without interaction. They will simply give up. Anything you say in the morning can change by the afternoon, so why pay attention at all?

7.10.14. Just One of Those Days

We've all had this experience, either for ourselves or from our employees. If you say, "I'll call it a loss and do better tomorrow," that's a forfeit and a waste of an opportunity to grow. When you find yourself facing a to-do list that you do not want to get into, find the one task or outside worry that is the real cause of your inertia.

Why do you need to do this thing? Connect the task to the good that will come of it.

Reflect on your fear. This feeling of boredom is actually based on fear. What are you afraid of? Identify it. Embrace that fear as an emotion that you feel for a reason, as part of yourself. When you name the monster, it loses power over you. Let your mind run on the fear if it wants to, but don't give it fuel. Eventually your pulse will slow and calm will settle.

Often the fear is that you will not be able to meet the ideal. You have a vision, and you know what quality is. But you must define reality, as it is. This is the task before you. Its reality does not have room for your vision. But it is better to be done than a to be a perfect castle-in-the-sky. Almost every deliverable can be improved more easily than made perfect on the first run.

Focus on why you are doing this task. Focus on the good. Whatever greatness or bad things that come from others' reactions to the results are not important at this point. Such non-reality indulgence makes a liability of the human imagination. Control it!

Embrace the suck.

The only origin for this phrase, as it is, that I could find was the United States military.

Define the task or the list of tasks as "suckish" and welcome that. It is days like this that make other days easy. "Hello, ugly thing to tackle. Welcome to my world. 'Now you're going to die.'"

This last line I took from Johnny Cash's A Boy Named Sue. Always a good song to keep in mind when embracing the suck.

Smile. Wallow in the thing you don't like. And move forward.

Take control of your time. Do one part of the task at a time, and move to the next. Set up a mini schedule and enjoy the small one-minute wins. Forbid yourself distractions. When you complete one mini-task, take a break of seconds to enjoy it by looking out the window or giving yourself a victory stretch. Then move on to the next part. Do not go to another task or email or other distraction until the whole thing is done.

This task, or day, as tedious or scary as it may seem, is just one more learning opportunity. You learn about your mind and your fears. You practice control of your imagination and of the minutes that make up your life. When the task is done, you will feel you had a "good day". Stop yourself from thinking, "I'm glad that's over." Instead, be consciously grateful for the suck, for the experience, for what you learned.

7.10.15. That's Interesting

Let your writers use their imaginations as much as possible. When you hear yourself saying, "We always do it this way," you know you are on the wrong track. Writers should know the style guide and procedures, and they should know when to break them. Do not make judgement calls when someone on your team comes with a new idea. If you can't see the benefits, ask, "Why?"

I can't tell how many times … yes, I can: three … that my manager asked me "why?" when I came up with some scheme that I thought was brilliant. The first time, it caught me off-guard and I couldn't answer. The second time, I was ready, but my manager showed me there was no ROI to my efforts. The third time, I realized just by saying the answer out loud that it was a stupid idea. Knowing when and how to ask "why" is a great skill for a manager to have.

If you ask, and still do not see the benefits, your next step depends on the setting.

- If the new idea comes up in a team meeting, a correct response is: "That's interesting. Let's discuss it further outside this meeting."

- If the idea comes up in a one-on-one meeting, do not be afraid to take a few minutes of silence, to seriously consider the idea. Ask more questions about the estimated effort and whether the results meet your success criteria.
If it is a go, start the Innovation procedure.
If it still is not a go for you, but it is for the writer, take some time: "I need to think about this. Let's meet tomorrow." You might change your mind. At least you give yourself time to consider how to communicate the stop of this project and retain the employee's will to use his imagination the next time.

7.10.16. Happiness is a New Challenge

When an employee has been on the same task for a long time and is bored with it, feel free to juggle the tasks. You can raise energy of multiple employees by switching products. Don't worry that you lost expertise. The improved energy will eventually improve productivity and quality. And now you have "high availability", to use a networking term: if one falls, the other is able to carry on with minimal down time of the project.

If an employee is interested in a non-writing task that is not on your list, such as programming or video creation, let them have it part of the time. Do not dump it on them. Come up with a plan that ends in real results. Use the engineering method to work it out. Let them schedule time during every week or month to work on this, while still meeting expectations for priority work.

7.10.17. Reviewing Performance

When you write a performance review, write about what the person did, not who you think the person is. Be consistent and objective in the written review. Make comments relevant for positive changes.

During the performance review discussion, ask "What do you want to do?" Make this a two-way dialog.

- Put the employee at ease.

- In a dialog, set out goals and priorities. Make sure the employee is committed with real engagement. Talk about things that can be changed. Problem solve together. Make sure the solutions will build confidence and self-improvement.

- Discuss actions for development as immediate action items. Agree on success criteria and measurements. Follow up later with coaching, status, and measurements. Be a partner in the quest for success.

- Accept feedback. Listen. Accept. Acknowledge. Ask for more information.

8. TWO CULTURES

In the book, The Two Cultures, C. P. Snow defends his Read Lecture. In the Canton Edition (1959), Snow explains why technology is an important component of the advancement of the human race. Snow explains how the educated can help technology fulfill its potential. In this, I find my higher purpose. If our writing is concise, relevant, consistent, easy to use, and in every way meets our definition of quality technical writing, we can make the world a better place.

A quote from The Two Cultures: "Most of our fellow human beings, for instance, are underfed and die before their time. In the crudest terms, *that* is the social condition. There is a moral trap which comes through the insight into man's loneliness: it tempts one to sit back, complacent in one's unique tragedy, and let the others go without a meal."

Can technical writing feed the hungry, heal the sick, bring people together, shelter all humans? Snow says we had, even in the 1950s, the technology to do all of this. It depended only on the "spread of the scientific revolution all over the world."

Technology *is* spreading. Snow said technology could save the human condition. But access to technology and ability to use it are different things. How can a kid in rural Africa use a smartphone to get a better education than his or her parents had?

We make sure the knowledge to best use technology is available. We teach.

Snow wrote: "...technology is the branch of human experience that people can learn with predictable results." We can reach the world, one piece of technology at a time. We write in a simple, controlled language that can be easily understood or well-translated by a machine. We carefully check our procedures for accuracy. We make sure the relevant documentation is easy to find.

A farmer in Argentina wants to feed his family. He reads on the Internet how he can make more money with flour made from the coffee bean left-overs. He can learn about this innovation because a technical writer explained the idea in language that Google Translate understands, with easy-to-follow steps, and with easy-to-understand graphics.

A woman in Chile buys medicine from a pharmacist. She can't afford the best, but she insists on getting something. She is given a packet of pills that haven't passed the US FDA. Her daughter asks the mother to wait a minute before she spends her money on the pills. The daughter looks up the medicine on her phone. She reads an article written by a pharmaceutical company tech writer. She does not need to translate it. The article is easy to understand and her English is good enough. This brand of medicine was pulled from the shelves in the USA because of the horrible side-effects. The daughter asks the pharmacist if he has a generic medicine mentioned in the article. With a few added pesos, she saves her mother from a painful death.

A group of young Ghana college graduates hears of a conference about water usage. They go to a session given by a man who developed a system to build wells from cheap, at-hand materials. They ask to learn how to make these wells. They are given a pamphlet in simple English. They start a company and bring their knowledge to the villages. They go through a hundred villages in a yearly circuit to maintain the wells. Each village is able to save enough during the year to pay the young entrepreneurs when they come by.

As a technical writer, I earn a salary. That is my profession. My passion, my purpose, is to use my skills and access to knowledge and experts, to make the world a better place.

> "[People of the future] may try to improve the quality of their lives,
> through an extension of their responsibilities, a deepening of the affection
> and the spirit, in a fashion which, though we can aim at it for ourselves
> and our own societies, we can only dimly perceive."
>
> —C.P. Snow

www.ingramcontent.com/pod-product-compliance
Lightning Source LLC
Chambersburg PA
CBHW030617220526
45463CB00004B/1315